Keeping Unusual Animals As Pets

 Sterling Publishing Co., Inc.   New York

Dedication

This book is dedicated to my beautiful wife, Veronica, for her love and support, and to my son, Nicholas, who will probably want a dog one day.

Acknowledgments

I would like to thank the following people for their support and encouragement: Veronica Hewitt, Nicholas Hewitt, Jerry and Carole Hewitt, Ruth Gamba, Dean Cains, Alex and Betsy Gamba, Eric Akaba and Trish Cannon, Sheila Barry, Jeff Arden, Jim Grizzell, Ed Kammer, Steve Kutcher, Rocky Lisa, Bruce Pierce for his assistance in selecting the right photographic equipment, Chris and Gene Roscher, Art Scott, John Shaw for his writings about close-up photography, and last, but not least, Iggy Hewitt.

Edited by Keith L. Schiffman

Library of Congress Cataloging-in-Publication Data

Hewitt, Jef, 1958–
    Keeping unusual animals as pets / by Jef Hewitt.
        p.   cm.
    Includes index.
    1. Amphibians as pets. 2. Reptiles as pets. 3. Invertebrates as pets. I. Title.
    SF459.A45H48   1990
    636.088′7—dc20   90-9986
                                                            CIP

10 9 8 7 6 5 4 3 2 1

First paperback edition published in 1991 by
Sterling Publishing Company, Inc.
387 Park Avenue South, New York, N.Y. 10016
© 1990 by Jef Hewitt
Distributed in Canada by Sterling Publishing
% Canadian Manda Group, P.O. Box 920, Station U
Toronto, Ontario, Canada M8Z 5P9
Distributed in Great Britain and Europe by Cassell PLC
Villiers House, 41/47 Strand, London WC2N 5JE, England
Distributed in Australia by Capricorn Ltd.
P.O. Box 665, Lane Cove, NSW 2066
Manufactured in the United States of America
All rights reserved

Sterling ISBN 0-8069-7278-5 Trade
            0-8069-7279-3 Paper

CONTENTS

Color section follows page 64

PREFACE

This book was written for those who are looking for a pet that most people wouldn't want, appreciate or understand. The following chapters were written to enable you to keep as a pet your choice of some strange creatures which will do well in captivity. I have written about animals that (from my personal experience) will live a relatively long time if you follow the guidelines that I suggest.

I *could* have included many other animals, but I didn't, because I wanted to cover just those animals that are worthwhile having as pets. After all, no matter how unusual an animal is, who wants to keep something that will just suffer and die in captivity long before its time?

Many of the creatures described here can not be found in any other pet book, or there is not enough written information available to keep these creatures in captivity with any success, or the animal's true value as a pet has been overlooked. If you're looking for the unique, then read on and enjoy, and I hope that by sharing my experiences and knowledge with you, you may have some of the fun that I've had.

THE FIRST STEPS

Why Keep an Unusual Pet?

The inevitable question that is asked when dealing with unusual animals is, "Why would anyone want to keep *that* as a pet?"

There are some good answers to this question. Most of the animals described here don't require much space, so even an apartment dweller can have one. Some of these animals require little care compared to the care a dog or cat might need. From personal experience, I can say that there are two *main* reasons why most people keep unusual creatures as pets.

First, I believe that keeping such pets is a function of individuality. In a world of dog, cat, and bird lovers, a man or woman who owns a pet scorpion is unique. An interesting animal in a person's home becomes a real conversation piece when guests are around. It is not uncommon for a teenager to seek out pets like the ones described here so that he or she can express his or her individuality. The "shock value" of these creatures to a teenager's parents is a definite plus from the teenager's point of view.

The second reason for keeping one of these animals as a pet is the pure fascination that some people have with nature, even in its most grotesque forms. For the true nature lover, there is beauty even in the most hideous-looking species, and the variety of nature's bizarre designs provides a never-ending source of fascination.

Ectotherms

All of the animals described in this book are *ectotherms.* Their internal body temperatures are significantly affected by the temperature of the external environment. To further understand this phenomenon, let us look first at humans.

We are *endotherms.* Our bodies self-regulate our internal temperature. The friction created by the blood circulating through our veins warms us, and we sweat to cool ourselves. Our skin insulates us against the environment. Although extremes in temperature can affect us, our internal body temperatures usually stay the same, in spite of external temperature variations.

Endotherm means that most of our temperature regulation comes from *within* us. Other endotherms include the rest of the mammals, as well as all of the birds. Feathers and fur help to provide insulation, while behaviors like panting can cool an endotherm.

Invertebrates, amphibians, and reptiles rely on the environment *outside* their bodies to help regulate their internal temperatures. This is what *ectotherm* means.

To see how ectotherms use the external environment to regulate their internal temperatures, let's look at the blue-tongued skink, *Tiliqua scincoides.* As the sun rises, the blue-tongued skink comes out to bask. This helps to warm the lizard and to get it going for the day's activities. If the temperature gets too high as the day progresses, the blue-tongued skink retreats to the shade or to a burrow to escape the heat. If the temperature begins to drop, the lizard may retreat to a burrow, where it will become dormant as its metabolism slows down. Like all ectotherms, the blue-tongued skink lives within the limits created by the temperatures in its environment.

Endotherms have been able to adapt to extremes in cold that would kill most ectotherms. On the other hand, because ectotherms do not expend much internal energy to maintain their metabolisms, they have been able to adapt to environments that do not have much available food, or to environments that have very high temperatures, like deserts.

Ectotherms (like reptiles) are sometimes referred to as "cold-blooded." This is a misnomer. Snakes, for example, are not "cold-blooded." They merely use the external environment to warm or to cool themselves. In fact, because many snakes will bask or stay in a warm place, they usually feel warm to the touch, rather than cold.

Some ectotherms are capable of raising their body temperatures by physiological means. For example, the females of certain python species are known to coil their bodies around their eggs, and by rhythmically contracting their muscles, they generate heat to help keep the eggs warm. With these facts in mind, you can see why the term ectotherm is preferable to "cold-blooded."

Many ectotherms remain inactive until a temperature is reached that is optimal for them. Because of the ectotherm's dependence on its external temperature, it is vital that you pay close attention to the temperature guidelines given in this book. Most ectotherms can be killed quickly by extremes of heat or cold, and some are less tolerant of variations than others. Be sure to use common sense and a thermometer, and you shouldn't have any problems.

Ethics, the Pet Trade and the Law

There are those who would argue that all of the animals described in this book should be classified as "wild," and that the keeping of them by private citizens should be made illegal. Indeed, some localities have passed laws that do just that. Most of these laws target snakes, but some laws have broader definitions of prohibited wild animals.

Many times, fear is the motivation for the passage of such laws. People fear what they do not understand. Don't flaunt the fact that you have one of the animals described in this book. Even if there are no laws in your locality against the keeping of these animals, you still have an obligation not to scare people. Some people will *never* learn to appreciate these creatures. If you choose to appear in public with your *large* snake, for example, you may further damage the animal's already unsavory reputation.

Other than the local laws that govern the keeping of animals by private citizens, there are also other government regulations that affect the pet trade. These laws are designed primarily to prevent animal abuse and to protect natural environments. Australia bans the export of its native wildlife; the cockatoo is one example. California's laws prohibit the sale of its native animals; the southern alligator lizard is one example. Most of these laws have been well thought out, but many of them are difficult to enforce; there are just too many pet owners, and too few wardens.

Those who have had extensive experience in the pet trade will tell you that, for the most part, it is a good industry. Like any other business, it has its share of unscrupulous individuals who have no regard for the law, or for anything other than making quick money. These people hurt the industry as a whole.

There are many "horror" stories about the bad side of the pet trade. Most of these stories concern the keeping of animals under cruel conditions that include over-crowding, starvation and a lack of cleanliness. These stories also often involve another illegal activity, such as the smuggling of protected or endangered animals. *Legitimate* pet dealers will do their best to obey the laws and to keep their animals healthy.

These "horror" stories are almost never about reptiles or amphibians. There are probably two reasons for this. First, most people don't know how ectotherms should be kept, so they don't recognize cruel conditions. Second, most people have very little sympathy for the types of ectothermic animals described in this book.

To illustrate the lack of compassion for these animals, let me describe vicious events called rattlesnake round-ups, sometimes held in Texas, Florida, Georgia, and Arizona. During these "festivals," people go into the wild and capture as many rattlesnakes as they can. They then place the rattlesnakes in large open pens that have no protection from the sun. As they dump the snakes into the pen, they prod and kick them. Some of the snakes had been caught in advance, and they are not fed for weeks before the events.

As part of these events, "professional" rattlesnake handlers enter the snake pens, and then show their skill at catching the snakes without being bitten. They remove the snakes, most of which are disoriented, injured and starving, one at a time, to be killed before the paying spectators.

After each snake is killed, it is skinned, its meat is cooked and sold, and its skin is made into hats, belts and other items.

If the animals were squirrels, or chipmunks, could you imagine such "festivals" being allowed to take place? Such cruel activities would never be allowed in a civilized country without public outcry for the animals. Don't you think that such "festivals" would be outlawed immediately?

Nothing more needs to be said about man's lack of compassion for most ectotherms.

As a prospective unusual-pet owner, you should be aware of all of the facts concerning these animals. Always know the laws governing the pet trade in your area. Seek to obey them and to encourage others to do the same. If the laws seem unreasonable, find out how you can act to change them.

Patronize pet stores that keep their animals in good condition. If you find a pet shop where a reptile, amphibian or invertebrate is suffering needlessly, notify the management in a helpful and courteous manner. This is usually sufficient to get things changed for the better.

Do not purchase animals that you know should not be sold. These include protected, endangered, or smuggled animals. There are plenty of animal species available as pets that are not threatened in the wild or that are captive-bred in large numbers.

It is quite possible that any of the animals described in this book could someday become endangered in the wild. If this happens, don't try to get the animal unless it is being captive-bred successfully, and you can confirm the origin of your animal.

Act in an ethical manner. Do what you can to improve the image of these ectothermic animals. Don't scare others, and don't parade your pet needlessly in public. On the other hand, in a controlled environment (such as a classroom or an exhibition), talking about your pet and displaying it can help to make people realize that these are living creatures which, like all other animals, deserve respect. You could even help to save an animal from extinction by influencing others to protect it.

Do not attempt to keep one of these animals if you are not serious about maintaining it properly. These are not domestic animals, like cats and dogs. If you are not prepared to give your unique pet the things it needs, it is better to leave it in the wild. If you seriously seek to develop an understanding of your animal, then by all means keep it, for your understanding of some aspect of such an animal may be the one thing that can help to save some of these creatures in the wild.

Choosing Your Unusual Pet

Selecting an animal can determine whether your experience with unusual pets is a positive or negative one.

Many people who attempt to buy an unusual pet are not armed with enough knowledge to ensure that their newly acquired pet is a good one. For the unlucky ones, this usually results in the pet's death. The person blames the animal or the hobby in general, when it was actually the fault of some other factor of which they were not aware.

If you plan to buy an animal, one of the most important things you must do to ensure the health of your animal is to develop a good relationship with a good pet store or private pet seller. The best way to achieve this is to visit the store or location and to talk to the other customers.

Don't be afraid to ask questions. Have the other customers' experiences with the store been good? Have their questions been answered satisfactorily? Were the employees willing to help? Did the animals they purchased from the store do well?

Next, examine the store or facility itself. Are the animals kept in clean conditions? Are the animals given proper care? As I stated previously, these are *living animals,* and if a store or dealer is not treating them well, then do your business elsewhere.

You will find that a reputable store or a private seller will be more than willing to answer your questions, since these suppliers will want your experience with these animals to be a good one. Any good supplier of unusual animals will not hesitate to allow you to watch your prospective pet eat, and this can be an important way to make sure that the animal you are buying has adjusted to captivity.

Occasionally there are specimens for sale in pet stores which have not yet adapted to captivity. Most stores have a designated "feeding time," and they will let you watch your prospective pet eat (before you buy it) so that you can make sure that it has adjusted to the extent that it (at least) eats well, and that it is free from disease.

Examine the animal closely. Are its eyes (if it's an animal whose eyes you can see) clear? Does the animal seem alert? Is it free from external parasites? Make sure that the animal does not appear too thin, and that all of its limbs (if it has limbs) are without infection.

You can ask the pet store owner or private seller to give you the names of other people who have had experience with the type of animal you are buying. Many areas have pet organizations or even herpetological or entomological societies where you can reach such people. Pet stores or local zoos will usually have information on these groups. These "clubs" are quite worthwhile since they can often provide you with sources for veterinary care, should the need arise.

If you buy an unusual animal, examine your source, check with the current customers, observe the animal you want, and if everything looks good, purchase your new pet and enjoy it.

Caring for Your Unusual Pet

No matter how odd these animals look, they are still living creatures with certain needs. Most of these animals can not cry out if they are hungry or if they're in pain. As their keeper, *you* must make sure that all of their needs are met. Although for most of these species care is relatively simple, it is still very important to do whatever is necessary to keep the animal healthy. If you are not prepared to follow through and maintain your pet properly, then for the animal's sake, don't keep it.

There is a basic care checklist that applies to all animals, and if you find an unusual animal that is not listed in this book, then you can still follow this checklist to ensure that your pet will stay healthy. Specific requirements are listed separately for each animal in the subsequent chapters.

The first requirement is food, and the main concerns are *what* and *how much* should the animal eat. As you will see, some animals eat only on a weekly basis, some need more frequent feedings, while others eat in cycles and fast for certain periods of time.

Some animals have very specific food requirements, while most need a variety of foods to provide them with proper nutrition. Use vitamin supplements to ensure this proper nutrition. Powdered bird vitamins or reptile vitamins containing vitamin D_3 and calcium are the best food additives for most of the species described here.

The second requirement is water. The amount of water required varies greatly from species to species. Some animals require a great deal of humidity in the air to maintain their health; some species require desertlike humidity.

The third requirement is warmth. Some animals need high temperatures to digest their food; some need to be kept cool to prevent excessive water loss and dehydration. One way to provide warmth is to use red incandescent light bulbs. They don't keep diurnal animals awake at night, nor do they interfere with the activity of nocturnal species.

One good arrangement is to mount the bulbs on a shelf that is closed on three sides, and then to place the container housing the animal on that shelf. The sides of the shelf will insulate the area; use two 25-watt red aquarium bulbs to provide a backup if one of the bulbs burns out (Illus. 1 and 2). The bulbs should be mounted so that there is plenty of clearance around each bulb (not less than 1½"). To prevent overheating, each bulb should not exceed 25 watts. This heating method also allows more than one tank to be heated on the same shelf.

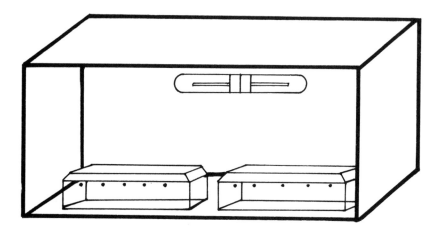

Illus. 1.   A shelf heated by two 25-watt aquarium-type bulbs mounted at the back of the tank. This lets you heat several plastic shoe boxes or tanks simultaneously. Allow clearance around the bulbs for air circulation. This will prevent overheating.

Never place a glass or plastic terrarium in the sunlight, even in a window; this raises the temperature quickly in the entire tank to a very high level and will not allow the animal to have a place to cool off. This could rapidly kill even the most heat-loving specimen.

Monitor the temperature by using a thermometer; just because the temperature feels "right" to a human doesn't mean that the temperature is correct for the animal. Inexpensive and easy-to-read plastic stick-on thermometers eliminate guesswork.

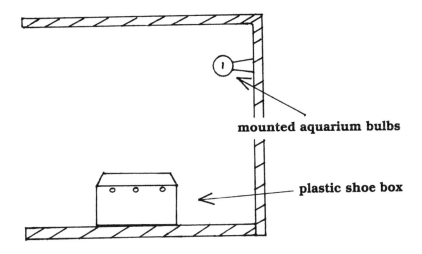

*Illus. 2.   This is a side cutaway view of the "heating shelf."
The front view is shown in Illus. 1.*

Light is another factor to consider. The amount of light may be crucial to some creatures' survival. Too little light (for some animals) can result in vitamin deficiencies, while too much light for others can cause stress. Some species need a definite day and night cycle to remain healthy.

The animal's general environment should be examined, as well. Pay particular attention to the container itself and the *substrate* (the flooring material). Is there a place for the animal to hide for security; are there objects on which to climb? The container must be secure from outside intruders (like cats or small children), and the internal environment should be appropriate to the specimen being kept.

Cleanliness is also an important factor to consider. An unclean environment can lead to infections and diseases. One of the best indicators of cleanliness is smell. If the container smells foul, cleaning is overdue.

All of these factors are intertwined. For many animals, if certain requirements are not met, they will not eat or drink. If the animal you are trying to keep as a pet is not eating, then ask yourself if all of its needs are being met. Is it too hot or too cold? Is it too wet or too dry? Is it being offered appropriate food? Can it hide and rest?

Once all of the animal's requirements are met, then usually it will do well. Certain specimens never adjust to captivity. In these cases, if the animal was caught in the wild, then it should be returned to its original home as soon as possible.

Animals should *never* be released in locales other than the one from which they originally came. The marine toad (*Bufo marinus*), for example, was released into the waterways of southern Florida to control pests in the sugarcane fields. The marine toad soon wiped out not only the pests, but native toad and frog species as well. History is filled with environmental disasters caused by careless individuals. Don't add to that history.

One last factor to consider is your home's environment. Once a man complained to me about having a lizard that wouldn't eat. After questioning him thoroughly, it seemed that he was providing the lizard with the proper heat, light, water and food. The lizard had a good-sized cage with a place to hide, and the only answer seemed to be that the lizard was diseased. When I asked the man one last question, he inadvertently revealed to me that the lizard's cage was on top of a large stereo speaker. The man liked to play loud music, and he never realized that the vibrations were causing the lizard stress. Once the terrarium was moved to a quieter place, the lizard began to eat and to do quite well.

Selecting a Veterinarian

When you have a cat or a dog, choosing a veterinarian is usually a simple matter. Asking other pet owners or looking through the telephone directory is usually enough to find a reputable "animal doctor" for your pet. When you look for a veterinarian to treat your pet Chinese crocodile lizard, your search may not be quite as easy.

There are a number of ways to search for a veterinarian to treat your unusual pet. The first way is to use the telephone directory. The Yellow Pages usually have advertisements that include descriptions of the types of animals that a particular veterinarian can treat. Be aware that the term "small animals" in a veterinary advertisement usually refers to small mammals and birds, and does not always refer to small ectothermic animals. If these ads state that a veterinarian treats reptiles and "exotics," the veterinarian can probably help you with amphibians and other ectotherms as well. Call the veterinarian's office to find out if he or she can deal with the type of animal you have. If a veterinarian tries to administer treatment to animals about which he or she does not have much knowledge, the kind of treatment can be totally inappropriate and could even kill your animal.

Because most veterinarians are conscientious professionals, if they are not capable of treating your pet, they can usually refer you to another veterinarian who is, and most will not hesitate to do so.

Another important source of veterinary referrals is a pet store that sells unusual animals. Many of these stores will have files of names of veterinarians with whom their customers have had good experiences. This is a good way to find someone who is capable of treating a particular type of animal.

A third source for veterinary referrals is to ask people who have unusual pets like yours. These people can be found through local herpetological societies, as well as through pet stores that deal in unusual animals. These people can usually give you helpful information. When it comes to dealing with a particular type of animal, they usually know who the best veterinarians are.

It is a good idea to find a veterinarian for your unusual pet before the animal actually needs one. Unlike cats or dogs, unusual pets do not usually require regular veterinary treatment, but it is important to have a good veterinarian in mind, just in case an emergency arises. Write down the telephone number and address of the veterinarian you find, and keep the information handy.

If your animals seem to be ill, there are several guidelines that you should follow. Don't wait too long to take action. Usually, some temperature or humidity adjustment is sufficient to remedy the problem.

If your animal does not seem to respond to these adjustments, check with your pet store or with other people who have the same type of pet. They might be able to suggest some solution to the problem.

If the problem persists, or you're uncomfortable administering a recommended home treatment, then contact the veterinarian that you have found. He or she may be able to advise you, or an office visit could be required. If your veterinarian's recommendation does not satisfy you, do not hesitate to seek a second opinion to ease your mind.

The animals described in this book are usually quite hardy. When given the proper treatment and captive conditions, their powers of recovery are quite amazing.

Very little is known about some of the animals found in this book (scorpions, for example). You may not be able to find anyone who can treat them. In this case, you'll just have to use your better judgment when treating them. Make an effort to record your actions, because you may discover a successful technique that you can share with others.

1
INVERTEBRATES

Crickets

Some people keep crickets as pets because they believe that their presence in a home brings good luck. I have included them in this book (Illus. 3 and 4) because they make excellent food for most of the insect eaters that are described in this book.

Crickets' nutritive value can be further increased by "dusting" them with a vitamin powder (a bird or reptile vitamin supplement) before you give them to another animal as food. The vitamin powder should contain calcium—this is one element that the cricket's body lacks.

Whether you are keeping the crickets for use as a food source or just for luck, they require a secure container. Many oriental gift shops sell small cricket cages. You could also use a plastic trash can. Place some cardboard or a used egg carton on the bottom for use as a hiding place, and the crickets will not attempt to climb out.

The worst misconception about crickets is that they need only a diet of vegetable matter. Crickets need a large percentage of protein, and they should be fed some kind of leftover meat. The best food for crickets (from my experience) is a "mash" diet that is made for young chickens. This mash is readily available at most "feed stores," and at many pet stores. To feed the crickets, place some mash in a dish and then keep refilling the dish.

Illus. 3.   A cricket, Acheta domestica, *crawls over a rock. It uses its antennae to explore the environment.*

Illus. 4.   A pair of crickets. The female (larger than the male) has a long ovipositor for egg-laying.

Provide water in a dish filled either with wet cotton or with paper towels that are kept moist. This method is most effective because crickets may drown in standing water.

As long as you maintain a heat level above 65°F, your crickets should be fine. If you want to breed crickets, you must provide some moist soil in which the females can lay their eggs, and you must keep the temperature above 80°F for steady breeding and rapid growth to take place.

Millipedes

There are over 10,000 different named species of this member of the class Diplopoda, and they can be found in many places, from backyards to deserts, usually in the piles of decaying plant matter which the millipede eats.

Millipedes can be distinguished from centipedes since millipedes possess *two* pairs of legs for each body segment, while centipedes only have one pair of legs per body segment (Illus. 5).

The larger species of millipedes, from tropical areas of Southeast Asia, Central America and Africa are quite impressive in appearance, having body lengths of over 6 inches. They are all relatively easy to keep as pets.

A plastic shoe box with some small holes drilled in it for ventilation works well for these species, since the box retains the humidity that the millipedes require. Several specimens can be kept together, if desired.

Place this box on a shelf heated with red light bulbs (see the section of the introduction that deals with caring for your unusual pet) to keep the environment around 80°F. This is the ideal temperature for most of these tropical species.

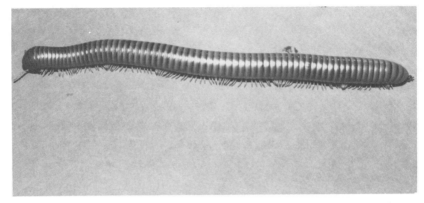

Illus. 5. This species of millipede uses a "wave" action of its legs as it walks.

Design the captive environment by keeping in mind the natural environment from which the specimen came. A flooring material of potting soil suits almost all specimens, and the small species usually will use a piece of cardboard or bark for a hiding place. A small water dish with a stone inside (for easy access) will fulfill the water requirements of most species.

Vegetable matter is the best diet for most millipedes; provide the millipede with a steady supply of fresh romaine lettuce, escarole or summer squash. These vegetables can be occasionally dusted lightly with a supplemental vitamin powder. Never use iceberg or head lettuce, because they consist mostly of cellulose and water, and lack sufficient vitamins and minerals necessary for proper nutrition.

Most millipedes have two defense mechanisms. First, they roll themselves into a coil, and their hard exoskeleton becomes the only surface exposed to attack. Second, many species of millipedes secrete a toxin. Some species have a very potent excretion. Although the toxin is not dangerous unless it gets into a cut or into mucous membranes (like the mouth or the eyes), and although the toxicity of the poison varies greatly from species to species (with many being harmless), you should wash your hands before you touch your food, or your mouth, or your eyes after handling a millipede. Never handle a millipede if you have an open cut on your hand.

Millipedes are interesting creatures, especially since they walk with their legs moving in "waves" as they travel. Most species of millipedes do not bite in self-defense, so handling them provides an unusual sensation as they walk.

One of nature's unique creations, millipedes make intriguing pets.

Centipedes

Centipedes are members of the class Chilopoda; there are 2,500 species. Centipedes can be distinguished from millipedes since they possess only *one* pair of legs per body segment (Illus. 6) as opposed to the millipede's two. Centipedes tend to move much more rapidly than the millipedes do.

Centipedes feed on living animals (for the most part) rather than on plant matter, and their speed enables them to capture prey.

The centipede has a pair of large *maxillipeds* which are pincerlike structures under its head. These maxillipeds are used to catch and hold prey. At the base of each of these hollow "fangs" is a venom duct that allows the centipede to inject poison when it bites.

The centipede's swiftness and biting ability make it impossible to grasp with your bare hands. It is best captured and transported in a plastic dish with a lid on it, since it will bite readily and quickly.

The centipede is a fascinating animal to observe, and will usually do well in a plastic shoe box or any similar,

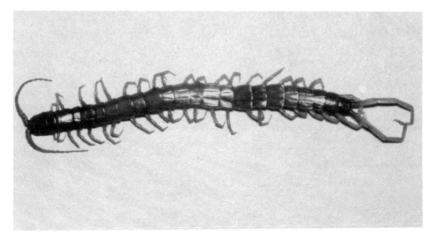

Illus. 6. The centipede is easy to recognize since it has only one pair of legs per body segment.

secure container. Fill the container with a sandy soil or moist soil base, depending on the area from which the centipede came. These animals can usually be found under rocks or debris, so they should also be provided with an appropriate hiding place.

A moderately stable room temperature is fine for these hardy creatures, and a small water dish with a stone in it (for easy access) will suit them well.

In the wild, the various species of centipedes eat everything from earthworms and insects to scorpions. They will also eat each other, so they should be kept singly.

The centipede's venom serves to subdue its prey, so use extreme caution when changing its water, or moving or catching a specimen. Although a centipede's bite is said to be merely painful, it would not be wise to take any risks, especially if you are sensitive to insect bites or stings.

In captivity, crickets and other wild-caught insects make an adequate diet for the centipede. Many centipedes will live for years with a minimum amount of attention. Centipedes are able to survive for long periods of time without food, and usually a few insects each week are sufficient for these amazing creatures.

Members of the genus *Scolopendra* range over most of the United States, and some species found in the South and Southwest are commonly over 6 inches long.

If you don't mind not being able to handle your pet, and you want an animal that exhibits amazing speed and efficiency, then a centipede is ideal for you.

Scorpions

Scorpions must be some of the most formidable-looking animals in nature. They first appeared during the Silurian period some 450 million years ago. The *pedipalps* (pincers, or claws, as they are commonly called), armor-plated body, and venomous *aculeus* (stinger) have ensured the scorpion's survival in its hostile environment.

The scorpion's four pairs of clawed walking legs allow it to climb over just about any surface, and at any angle. Most scorpions possess a pair of eyes on the top of their heads (in the center) that appear to be two tiny bumps, along with 2 to 5 light-sensing *ocelli* on each side of the front of the upper body.

Some cave-dwelling species do not have eyes at all, but rely heavily on the long hairs (*trichobothria*) that project from the body and limbs of all scorpions. These hairs are quite sensitive to touch and aid the animal in sensing prey, mates, and enemies.

Scorpions also possess a pair of *pectines* (appendages that project from the breastplate) that form a "V" on the underside of their bodies. Although scientists are unsure of the actual function of these "arms," they appear to be used to actively touch the ground as the scorpion walks, giving the scorpion some kind of sensory input.

The mother scorpion can bear as many as 95 live young in one birth. The mother carries them on her back, thereby protecting them until they are large enough to fend for themselves.

The scorpion's reputation as a deadly animal can not be applied as a general rule, because out of 1,500 species, only about 50 are truly dangerous to man, and all of those 50 belong to one family, the Buthidae.

It has been suggested that the potency of a scorpion's venom can be judged by the size of the scorpion's pedipalps. Scorpion species possessing large, thick pedipalps would have a weaker venom than those species having more slender pedipalps. This stands to reason, since a scorpion with a stronger venom would not need other powerful means to defend itself and to capture prey. However, don't risk your life on the accuracy of this theory.

There is one scorpion species, *Pandinus imperator*, the African emperor scorpion (Illus. 7), which can be found in pet stores. The African emperor scorpion is one of the largest species of scorpions, with its body reaching a length of over 6 inches from its head to its stinger. Despite its menacing black body and incredible size, the venom of the African emperor scorpion is comparatively weak.

I have been stung by the African emperor scorpion, and I did not feel anything more painful than a pinprick. Although most specimens become "tame" enough to handle, I would not recommend handling them, since the pedipalps can inflict a painful pinch. Someone with known allergies to insect bites or stings should be particularly careful.

The African emperor scorpion can be kept for years in a plastic shoe box (or similar container) with small holes drilled in the box for ventilation.

By using a shoe box with a small water dish inside and by using potting soil as a substrate, the humidity level in the box can be kept high. This is important, since the African emperor scorpion comes from areas with high moisture levels. If the African emperor scorpion is kept in a dry or desert environment, it usually does not live very long.

A diet of crickets and other insects dusted with supplemental vitamins will nourish this animal well. A stone protruding from the water dish will prevent insect food from drowning in the dish and then polluting the scorpion's water supply. The stone will also provide the scorpion with easy access to the water.

Scorpions survive on relatively small amounts of food. Six large crickets each week will provide plenty of food for an adult scorpion.

Light is not necessary to sustain scorpions since they are nocturnal animals. The African emperor scorpion seems to benefit from a temperature between 75° and 85°F.

Illus. 7. The African emperor scorpion is one of the largest species of scorpion. It can be recognized by its large pedipalps *(pincers).*

Although African emperor scorpions can be kept in groups, it is usually best to keep them separately, since specimens will fight, especially if they are not introduced to the new captive environment at the same time.

Use a plastic dish with a secure lid to move or capture any scorpion, especially when the specimen being handled is unknown.

Members of another scorpion genus, *Hadrurus,* such as *Hadrurus hirsutus,* the Mojave giant scorpion, from the southwestern United States, do well in captivity. Use a mixture of potting soil and sand as a substrate. Use a screen cover on their container for ventilation (rather than a plastic lid) so that the humidity remains at a moderate level.

Whatever scorpion you decide to keep (I do not recommend any dangerous species), exercise caution to avoid being stung or pinched. The scorpion is a creature that garners a reaction from even the most fearless individual. It is truly a fascinating member of the animal kingdom to keep as a pet.

Vinegarones

Perhaps you've never heard of the vinegarone—that's one reason why this creature makes such an interesting pet.

Vinegarones are sometimes called whip scorpions, although they are from a completely different order, the Uropygi. They do look like scorpions, with their large pedipalps (pincers) and four pairs of legs, but the first pair of legs is long and slender and is used as "feelers," while the tail is very thin and does not end with a stinger (Illus. 8).

The tail is interesting because the vinegarone uses it to spray an acetic acid solution for defense. This solution is harmless to man, and it smells like vinegar, hence the name vinegarone.

The vinegarone produces no toxins and is, therefore, harmless. The sizeable pedipalps, however, can inflict a strong pinch, especially the pedipalps of one species, *Mastigoproctus giganteus*, which can reach an overall length of over 5 inches.

A vinegarone can be maintained in a plastic shoe box or similar container with a substrate of sandy soil. For ventilation, use a plastic lid with small holes in it, or a screen cover.

A small water dish with a stone protruding from it will allow the vinegarone easy access and keep the insects used as the vinegarone's food from drowning.

Introduce several large crickets (dusted with vitamin powder) every week to provide a healthy diet. For variety, offer different insects occasionally along with the crickets.

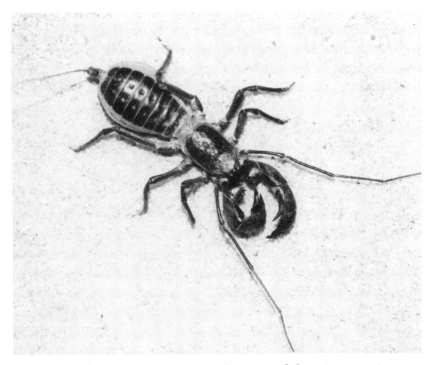

Illus. 8.   The vinegarone may be one of the strangest creatures in nature. This is the species Mastigoproctus giganteus.

Since the vinegarone is nocturnal, it does not need artificial light—it prefers darkness. The vinegarone seems to be best suited to temperatures between 70° and 80°F, so room temperature is usually acceptable, as long as extremes are avoided.

The vinegarone species *Mastigoproctus giganteus* is usually found under wood or debris in its natural habitat in the southwestern United States. Occasionally a vinegarone ends up for sale in a pet store.

Regardless of where you found it, if you have acquired a vinegarone, you'll have a unique pet.

Praying mantis

The praying mantis is a unique animal. It is one of the few insects that is respected by man. In fact, in the midwestern United States, most people teach their children to avoid killing the praying mantis.

The praying mantis has earned such respect because it is a voracious predator. It will eat large numbers of household and agricultural insect pests. In fact, praying mantis egg cases are sold in gardening shops and in plant nurseries. These egg cases usually come with instructions on how to hang them in your garden and how to facilitate incubation.

Unfortunately, the praying mantis does not limit its appetite to harmful insects, and it will readily eat its own kind. For this reason, it has been abandoned as a pest controller in commercial agriculture. Its cannibalistic tendencies keep its numbers too low to have a significant large-scale effect on pests.

This doesn't mean that you shouldn't try to keep a few in your garden. They might eliminate some nasty bug that has its eyes on your tomatoes.

The praying mantis is a member of the family Mantidae. There are a number of species that are referred to as a praying mantis, including the European mantid, *Mantis religiosa*, the Chinese mantid, *Tenodera aridifolia*, and the Carolina mantid, *Stagmomantis carolina*.

Despite their common names, many species of praying mantis are found in areas to which they are not

Illus. 9. The lightning speed and powerful jaws of the praying mantis can make quick work of most insects.

native. They have been introduced by man both delib-
erately and accidentally. The three previously mentioned species all can be found in the United States.

The praying mantis gets its name from its forelegs,
which resemble a pair of arms held as if in prayer (Illus.
9). These forelegs are equipped with spikes which make
catching, holding, and killing prey quite easy. The action of these forelegs is amazingly fast. Studies have
timed the strike of the forelegs at less than 0.1 of one
second—fast enough to catch a resting fly before it has
a chance to take flight.

The praying mantis is well camouflaged to aid its am-
bushing abilities. It remains motionless on a branch or
twig, and it blends in perfectly with the foliage, until
some small, hapless creature passes too close to it.

In the wild, the praying mantis will eat not only insects, arachnids and its own kind, but it has also been
known to eat small lizards, frogs and even humming-
birds. The female of the species (which is larger than
the male) has been known to eat the male before he can
finish mating with her.

The head of the praying mantis is mounted on an
unusually flexible neck which allows it to observe a wide
area. The jaws are particularly powerful, and they are
more than capable of cutting through the best-armored
insects, including beetles and wasps.

Many people have kept the praying mantis in a jar
with holes in the jar's lid for ventilation. A small aquarium with a screen cover makes a better home, because
it provides the animal with more room to move.

A substrate of aquarium gravel, potting soil, or even
newspaper is acceptable. Real or imitation plants or
branches should be added to allow the mantis to climb
freely.

A small water dish with a stone inside (for ease of
access) is recommended for most species. When design-
ing the environment, you should consider the area from
which your particular specimen comes. Make an effort

to closely duplicate the humidity and temperature levels of the original habitat.

Most species must be protected from chills. Red 25-watt aquarium bulbs and a thermometer will ensure the stability of the temperature of the vivarium.

If you provide a proper environment, the praying mantis will eat readily in captivity. Crickets make a fine diet. To ensure a sufficient amount of food, base the size and number of crickets on the size of your praying mantis.

It is particularly interesting to watch a praying mantis catch a fly, because of the speed with which the attack is launched. If you introduce flies or any other wild-caught insects as food, you must be sure that they are not contaminated by pesticides, or your praying mantis will quickly succumb to the effects of such poisons.

As was previously mentioned, the praying mantis will readily attack its own kind. With this in mind, keep only one praying mantis per tank. Because of its belligerence, the praying mantis is used for staged fights in some parts of the world, and wagers are made on chosen opponents in the same manner as cockfights.

If you are gentle, you can handle the praying mantis. Be aware of the sharp spikes on the forelegs. Some specimens are strong enough to pierce the skin with those spikes. Also, don't forget that some specimens can fly.

If you attempt to hatch a mantid egg case, remember that (in some cases), hundreds of tiny mantids can hatch from one egg case. For this reason, if you hatch the egg case indoors, you must use a fine mesh cover and a secure tank to prevent escapes, and if you hatch the egg case in your garden, be sure to close your windows, since the mantids may fly back into your home.

Newly hatched mantids will quickly start to feed on anything that they can catch, including each other, but you will still be left with several large praying mantes which can then be separated.

In the wild, most praying mantes will die at the first drop in temperature. Maintain the temperature at the appropriate level for your specimen, and you may be able to keep it for a few years.

Some species of the praying mantis are imported for sale to the pet trade. A specimen from a foreign area or country should *never* be released into the wild. This could have severe negative effects on the environment, so behave responsibly if you acquire a nonnative specimen.

Whether you keep an exotic mantid in a vivarium or release a native mantis into your garden, the praying mantis makes an educational and fascinating pet.

2
CAECILIANS

Caecilians

The caecilian is an amphibian not familiar to most people. All of the caecilian species are found (in the wild) close to the equator.

Ranging in size from 2½ inches to almost 5 feet, the caecilian closely resembles an earthworm, with its cylindrical, legless body and moist skin (Illus. 10). There are two types of this strange creature which may be found in pet stores.

Typical of the first type of caecilian is the yellow striped caecilian species, *Icthyophis glutinosis,* a terrestrial animal from Southeast Asia. As its name implies, it has two bright yellow stripes, one running down each side of its dark body.

The yellow striped caecilian averages about 10 inches in length, and spends its time burrowing in loose topsoil, so moist potting soil serves as a good substrate.

An aquarium with a screen cover or a plastic shoe box with holes for ventilation make ideal containers. The plastic container is usually preferable, because it helps to maintain the high humidity that this creature needs. The soil should always be moist but never too "muddy" in its consistency. A good rule is to keep the soil as damp as you would for a plant that needs a lot of moisture. With these humid conditions, exercise care to keep

Illus. 10.   This is Gymnopis multiplicata, *a terrestrial cae-cilian from South America. It is occasionally found on the pet trade.*

the soil from spoiling. If the caecilian's environment begins to smell foul, change the soil immediately. A ½-inch layer of aquarium carbon below the potting soil can help to prevent soil spoilage.

This species of caecilian will usually function well at a stable room temperature around 70°F. If you must heat the container, use great caution to prevent overheating (above 80°F should be avoided) and the loss of the humidity that is vital to keep this amphibian's skin moist. The caecilian's skin must remain moist to aid its respiration.

This type of caecilian eats earthworms, and surprisingly, it is fast enough to catch small crickets. A steady supply of a variety of these two food items will usually be consumed at night, since this is a primarily nocturnal animal with small, almost invisible eyes. As you would expect with such a nocturnal creature, artificial light is unnecessary.

These are truly odd creatures, and some other terrestrial species may be available at pet stores. Most of the species will stay buried or hidden much of the time. Handling them for any length of time is not recommended, since their skin is quite delicate and should remain moist. Nevertheless, the terrestrial caecilian is a great pet for the die-hard individual.

The *Typhlonectid* species of caecilian, such as *Typhlonectes compressicauda*, is the second type of caecilian. These can be found in aquarium stores, because (unlike those previously mentioned), these caecilians are aquatic.

Aquatic caecilians are often erroneously called "rubber eels" or "blue eels," and average about 12 inches in length, with a uniform dark blue cast to their skin and a body diameter somewhat thicker than a pencil.

It is unfortunate that these caecilians are often sold in aquarium stores as occupants of community aquariums. Many times this causes the animal to suffer, since aggressive types of fish in the same tank will tend to nip the caecilian's delicate skin and cause open wounds to develop. These species are best kept in a tank without fish, or only with those species of fish that do not bite or attack.

The aquatic caecilian thrives at a temperature between 68° and 75°F, so it is important not to house it in a tank which is kept at a high temperature; this can rapidly kill it.

Like the terrestrial species, the aquatic species also like to burrow in the earth, so the aquarium substrate should be of soft soil. If you use gravel, then provide some rock caves in which the caecilian can hide, since this animal does not like a lot of light.

In all other respects, the aquarium can be set up just like one that you'd use for fish, with a heater to keep the temperature stable, a filter to keep the water clean, and any other decorations which you might find appropriate and attractive.

The aquatic caecilian will eat a variety of foods, including aquatic worms such as the tubifex, soft-bodied insects such as waxworm moth larvae (*Galleria mellonella*), and earthworms. Some specimens will even learn to take food from your hand. Feeding the caecilian every other day will usually keep it healthy.

Occasionally, a pregnant female caecilian makes it into a home tank. Watching these animals give birth underwater to one or two young is an amazing sight. Separate the young from the parents. Although cannibalism is not confirmed among these animals, it is best not to take chances.

In general, caecilians are fascinating amphibians. Some exhibit peculiar adaptations, such as a pair of tentacles which protrude from the head. These tentacles are unique. Although they are used to detect prey in the soil, they have evolved from the structures of the eye. Muscles and glands that are associated with the eye in other vertebrates move and moisten these tentacles.

There are other species of caecilians which sometimes become available in pet stores, including terrestrial species that grow up to 5 feet in length. Whichever kind of caecilian you choose, you'll have a pet which most people have never seen before.

3
SALAMANDERS

Chinese emperor newt

The Chinese emperor newt (Illus. 11), *Tylototriton verrucosus,* is also called the rough-skinned, mandarin, or alligator newt, and is quite attractive, having raised orange spots on each side of its dark-skinned body. This salamander is native to China and Southeast Asia, and can reach an overall length of up to 8 inches.

A container such as a plastic shoe box with holes for ventilation with some potting soil for substrate will provide a good home. Some damp moss or a piece of wood or bark for a hiding place and a small water dish with a stone in it (for easy access) will complete the setup. In any environment where the humidity is high, exercise care to prevent decay and contamination. Your sense of smell is the best way to check this.

To provide the Chinese emperor newt with a larger and more attractive environment, an aquarium can also be used as a home. You must use a glass cover to retain humidity. The temperature should be maintained between 60° and 70°F. Heat should be avoided.

For food, a regular supply of crickets (dusted with vitamin powder), earthworms, waxworms *(Galleria melonella),* and some other insects for variety, will keep the Chinese emperor newt healthy and allow it to live in captivity for years.

Illus. 11.   The Chinese emperor newt, Tylototriton verrucosus, *is native to China and Southeast Asia.*

Remove excess food from the animal's home to ensure
that the animals you're using as food don't die and pol-
lute the Chinese emperor newt's environment.

The Chinese emperor newt has a defense common to
salamanders; it possesses a parotid gland on each side
of the top of its head. These glands secrete a toxin when
the animal is disturbed. This toxin deters predators,
and because of this defense mechanism, wash your
hands after touching the Chinese emperor newt. Of
course, as with all amphibians, the Chinese emperor
newt's skin is relatively delicate, so a minimum amount
of handling is suggested.

Axolotl

The axolotl may be one of the most overlooked of the world's unusual pets. It is interesting to watch, unquestionably unique, and is usually available in several colors, from black to albino.

The axolotl is actually the larval stage of the Mexican salamander, *Ambystoma mexicanum.* It is native to Lake Xochimilco and to Lake Chalco, located southeast of Mexico City.

One thing that makes this animal interesting is that it is a *neotenic* animal; it can remain in its aquatic larval stage for its entire life and can actually reproduce while in that stage.

The axolotl can be maintained in an aquarium. A 10-gallon size is adequate for two specimens. Treat the aquarium in the same manner as you would for keeping fish. There should be gravel on the bottom of the aquarium to allow for the growth of the bacteria that break down the animal's waste. Provide a good-quality filter to keep the water as clean and as clear as possible.

The two best filters for this purpose are the *external* type that hangs from the back of the tank, and the *canister* type that actually sits inside the tank. Each of these filters has its advantages and disadvantages. Your local aquarium store is the best place to seek advice on types and brands, since that store will be your source for new cartridges and replacement parts.

Monitor the temperature of the water by using a thermometer. Maintain a range between 65° and 70°F. Although this is a relatively low temperature, provide an aquarium heater to maintain stability and to prevent extreme fluctuations when the room temperature may be prone to variation.

The aquarium water will have to be dechlorinated for the axolotl just as it does for fish. Your local aquarium store can advise you if there is any other water treatment that is needed in your area. A neutral (7) to slightly high pH is advisable.

Illus. 12.   The featherlike gills of the axolotl, Ambystoma mexicanum, *can be seen clearly in this photo. This particular specimen is known as a "white" variant.*

The axolotl reaches an average overall length of 6 inches. It is a curious-looking creature, with vertical ridges along the sides of its body, and a number of featherlike gills protruding from each side of its neck (Illus. 12). The axolotl uses these gills to breathe under water. The axolotl should *never* be kept in an aquarium with fish. No matter what type of fish they are, they will invariably nip and eat these gills, so axolotls are best kept only with their own kind.

To keep the axolotl in great condition, feed it a varied diet that includes commercial fish food sticks or turtle food sticks or turtle food tablets, crickets and other insects, freeze-dried shrimp (sometimes known as krill when sold in the aquarium store), earthworms, and even an occasional newborn mouse.

Feedings of twice to three times each week are ideal, depending upon the size and the bulk of the food items offered. It is not unusual for the axolotl to learn to take food from the keeper's hand.

Warming and a reduction in the water level will cause the axolotl to metamorphose into its adult stage. In this adult stage it usually lives for several years, if kept in the same manner as the Chinese emperor newt, but it is more desirable to maintain the axolotl in its larval aquatic form. It is not unusual for the animal to live for 25 years if kept in this larval aquatic form.

One last thing that makes the axolotl such a wonderful and unusual pet is that almost all specimens available as pets are bred in captivity, so the pet keeper does not have to worry that by purchasing his pet he has taken it from its native habitat.

Palm salamander

The palm salamander, *Bolitoglossa dofleini,* belongs to the family Plethodontidae, known as the "lungless" salamanders. As a group, they have no lungs; respiration takes place through the skin. For a lungless animal, the palm salamander is quite large and robust. The snout-to-vent length averages about 5 inches, with an additional 3 to 4 inches of tail (Illus. 13). Females are usually slightly larger. The palm salamander is usually some shade of grey or brown, with various nondescript markings.

The palm salamander is arboreal, and has webbed feet that are sticky enough to allow it to adhere to most surfaces. In its native habitat, the palm salamander

Illus. 13. The palm salamander, Bolitoglossa dofleini, *is an arboreal salamander. The flat, webbed, sticky feet aid this animal in climbing.*

climbs in vegetation in search of the insects and arthropods that make up its diet. It uses its extensible, mushroom-shaped tongue to capture its prey.

This amphibian's tail is prehensile, and is used in climbing. If attacked, the palm salamander's tail has fracture planes which allow the tail to break off. The tail then continues to wriggle, distracting the attacker's attention and allowing the salamander to escape. This behavior is more common among lizards than among salamanders.

The palm salamander can be kept adequately in a 10-gallon aquarium with a screen cover, but a 20-gallon size is suggested because it can accommodate taller branches and it allows the palm salamander to have more climbing room. The branches should be anchored sturdily, and nontoxic materials must be used. If you decide to use real plants, make sure they are also non-toxic. Your local nursery can help you select appropriate specimens.

To keep the humidity high, you should partially cover the screen top on the palm salamander's tank, using glass or plastic. High humidity is important to the health of this amphibian.

To construct the best substrate for this salamander's tank, first place a 1-inch deep layer of aquarium gravel on the bottom of the vivarium. Mix in some aquarium carbon with this gravel to help prevent spoilage. Pour water over the gravel so that the water just covers the gravel. A few larger stones and some moss can then be placed on top of the wet gravel, and the branches can then be added and braced against the side of the tank. This setup provides the proper level of humidity, and the palm salamander can use the moss for hiding.

Substrates of newspaper, coarse aquarium gravel or potting soil can also be used. Also add a large, flat, water bowl. If you use this kind of setup, be sure to monitor the humidity. Closely observe the cleanliness of the tank (no matter which type of substrate you use), since higher humidity always leads to a high level of bacterial activity and the chance of spoilage.

The palm salamander will fare nicely on a diet of small-to medium-sized crickets dusted with vitamin powder. Waxworms (*Galleria melonella*) make a good food supplement, as do other small insects, and an occasional earthworm. All dead, uneaten food animals should be removed from the palm salamander's environment to prevent decay and the resulting odors.

Remember that this animal comes from a cool environment. The temperature should stay between 60° and 70°F. High heat can kill a palm salamander.

The palm salamander is nocturnal, so artificial lighting is unnecessary. If you want to have light over the tank, use a fluorescent bulb. To prevent excess heating of the vivarium, turn the light on only as needed.

Because of the palm salamander's relatively delicate skin and the fact that it may secrete toxins, keep handling to a minimum. If you do handle the palm salamander, avoid touching your eyes, nose or mouth. Always wash your hands after handling the palm salamander.

The salamander's detachable tail can be lost surprisingly easily. Handle the palm salamander carefully. Although the tail can grow back, you do not want it to lose its tail inadvertently.

If you do acquire a palm salamander, you will have a most unusual pet that will prove to be fascinating as it slowly climbs over the branches and the rocks in its environment. Keep the top of the vivarium covered—the palm salamander has no trouble climbing on glass.

4
FROGS AND TOADS

Tomato frog

The tomato frog, *Dyscophus antongili,* is an appropriately named amphibian. The adult frogs are generally very round and are sometimes a rich, red color.

This frog is from the family Microhylidae, and is native to Madagascar, where it lies among the ground cover, waiting for whatever prey it can cram into its mouth.

In captivity, the tomato frog can be kept in an aquarium that has a screen top. A substrate of newspaper, potting soil or aquarium gravel, along with a hiding place of moss or bark will complete the environment. Provide an elevated rock or piece of bark so that the frog has a dry place to sit. This prevents the formation of skin fungus.

A diet of large crickets or other insects powdered with vitamins, and an occasional newborn mouse will allow this frog to become quite robust. It may grow to over 4 inches in length from its snout to its vent.

Tomato frogs are currently being bred in captivity. Young tomato frogs become available from time to time. Their juvenile coloration is characterized by yellow and brown markings rather than the adult's red markings (Illus. 14).

Illus. 14.   Dyscophus antongili, *the tomato frog.*

If you acquire a young tomato frog, be sure to dust vitamin powder on the insects that you give it for food. Add some powdered calcium to the juveniles' water, since they appear to require relatively high amounts of this mineral in their diets.

Do not let the temperature of this frog's tank drop below 70°F. To prevent excessive drying of this animal's skin, avoid extreme heat above 80°F.

Keep handling to a minimum, since like most amphibians, this frog's skin is relatively delicate and the frog produces a toxin when disturbed.

The tomato frog's striking coloration, together with its round, piercing eyes make it quite a handsome creature. If provided the proper care, this frog will do well in captivity. For these reasons alone, it is obvious why this animal is a desirable unusual pet.

Rubber frog

This amphibian, *Phrynomerus bifasciatus*, is also known as the snake-necked crevice creeper and the red-backed walking frog. It earned these nicknames because it moves from one place to another in a very strange way. Compared to most other species, its back legs are shorter in relation to its body. This frog, from the family Microhylidae, usually walks, rather than jumps. As it walks, it uses its sticky toe pads to adhere to whatever surface it is traversing.

A native of central and southwestern Africa, the rubber frog has a muddy-brown to black body with two parallel red bands running down its back. It grows to a length of 2½ inches and for a frog, it has an unusually long neck (Illus. 15).

Illus. 15. The rubber frog, Phrynomerus bifasciatus, *differs from other frogs in that it tends to walk more than it jumps.*

In its native environment, the rubber frog inhabits termite mounds, vegetation and rotting logs. It eats tiny invertebrates that it finds there.

A glass aquarium with a screen cover, a substrate of coarse gravel, sturdy imitation or real plants (make sure to use nontoxic plants if you choose real ones), and pieces of bark for climbing and for hiding will provide a suitable environment for the rubber frog.

Maintain a temperature of 70° to 80°F. Provide a water dish, with a stone in it for easy entrance. A dish filled with moist sphagnum moss will give the rubber frog another place to rehydrate itself.

The rubber frog will thrive on a diet of small crickets and other tiny insects dusted with vitamin powder. Although it is not always found in pet stores, this fascinating frog is worth the search.

Chilean wide-mouthed frog

The Chilean wide-mouthed frog, *Caudiverbera caudiverbera,* is aptly named. It has a large mouth and an appetite to match. It will cram anything into its mouth that will fit, including mammals, birds, snakes, lizards, and even other frogs. This is a large frog, and it is not unusual for adults to reach a snout-to-vent length of over 9 inches (Illus. 16). At this size, it is obvious that this amphibian can take prey that is quite large.

The Chilean wide-mouthed frog is a member of the family Leptodactylidae. It has a robust body, short front legs and long back legs, which it keeps folded close to the body. Its color is a typical shade of green with some green or black mottling.

This frog is unique because of its size. It is certainly one of the largest frogs, and it uses its size to its advantage as it navigates the lakes and rivers of Chile and overpowers prey as large as itself.

The advantage of occupying these locales is obvious. Many animals come to the shores of a lake or a river to drink. Because of this the Chilean wide-mouthed frog is assured of a constant food supply. Other frogs are one of its main food items.

The Chilean wide-mouthed frog is primarily nocturnal. At the end of the night, it burrows backwards into loose dirt using its powerful rear legs to shovel the soil. It remains in this burrow until the end of the day, when it reemerges to search for prey.

Because of the Chilean wide-mouthed frog's large size, an adult specimen is best housed in a 40-gallon

Illus. 16.   The Chilean wide-mouthed frog, Caudiverbera caudiverbera, *is a huge amphibian that even eats other frogs. This young specimen is already over 4 inches in length.*

aquarium with a screen cover. The screen cover should be partially covered so that the humidity remains high.

A large, heavy water bowl is essential because this frog is highly aquatic. Provide a ramp made of either a piece of bark or a smooth rock to minimize splashing as the large specimens enter and leave the bowl. These ramps will also make access easier for young specimens.

A substrate of potting soil allows the Chilean wide-mouthed frog to burrow into the soil during the day. A lot of dirt will end up in the water bowl and you'll have to change the water frequently. Newspaper can also be used as a substrate, but you must provide a hiding place of cardboard, bark or plastic to give the animal a place where it can feel secure during the day.

Generally, a stable temperature between 65° and 75°F suits this frog well; room temperature is usually adequate. It is still a good idea to use a thermometer to monitor the temperature to ensure a proper level.

The Chilean wide-mouthed frog is bred in captivity. Look for a young captive-bred specimen, since wild-caught adults will sometimes refuse to eat. If you do get a wild-caught adult that refuses food, provide a large hiding place and put the tank in a quiet, low-traffic area, or cover the tank to induce feeding.

When the Chilean wide-mouthed frog does eat, it has a ravenous appetite, and it will eat anything—young or adult mice, fish, and even rats. Young specimens can be fed young mice, and crickets dusted with vitamin powder.

Young specimens should be fed a varied diet several times each week. Adults will usually do well on a weekly feeding of rodents. Some frogs will learn to take dead animals that are wiggled in front of them.

Specimens will usually eat more readily at night. Do not leave live rodents in the tank, since they can turn on the frog, and bite or wound it severely, especially if the frog is not hungry, and the vivarium is left unattended.

Because of its natural feeding habits, the Chilean wide-mouthed frog should never be kept with other

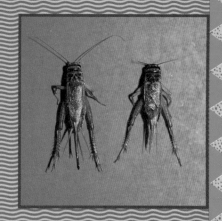

Top: a female scorpion with her young on her back. Above left: Chinese emperor newt. Above right: a male and a female cricket. Left: green toad.

Top: white axolotl. Middle: yellow-striped caecilian. Bottom: pancake tortoise.

B

Top left: rubber frog.
Above: carpet python.
Left: aquatic caecilian.
Below: blue-tongued
skink.

Above: Trans-Pecos rat snake. Left: Southeast Asian millipede. Below: Solomon Island giant skink.

D

frogs. To prevent cannibalism, specimens are best kept singly.

The Chilean wide-mouthed frog makes an excellent unusual pet. Given the proper conditions, this frog will usually live for years. They are robust animals, and their large size and highly positioned eyes give them an almost comical appearance. If you want a unique frog, the Chilean wide-mouthed frog makes a great choice.

Tropical clawed frog

The tropical clawed frog, *Xenopus tropicalis,* is a member of the family Pipidae. It is different from the frogs mentioned previously, since it is a totally aquatic amphibian, preferring to remain in the water for its entire life. The tropical clawed frog is native to West Africa, ranging from Guinea, south to Angola. It inhabits still and flowing waters in forests.

Xenopus laevis, the African clawed frog, is a more famous member of the genus. Like all members of the *Xenopus* genus, it too is aquatic. This species was widely used for pregnancy testing in the 1940s, but this was discontinued when cheaper, more effective means were developed. One of the adverse effects of this form of pregnancy testing was that many clawed frogs were released into the wild. Because of the hearty nature of these animals, they were able to survive in many areas of the world. In the 1940s, the African clawed frog was released in southern California, where it soon killed most of the game fish in the waterways around Los Angeles.

Because of this, many areas have legislation which not only forbids the sale and possession of *Xenopus laevis,* but forbids the sale and possession of all species of *Xenopus.* Check your local laws before you attempt to find and purchase *Xenopus tropicalis*. Make sure that it is legal to acquire a tropical clawed frog, and once you do, it is your responsibility to see that it is not released into the wild. The environmental impact could be disastrous.

This frog is well adapted to an aquatic existence. Its eyes are placed on top of its head, and it has a dorsoventrally flattened body which allows it to glide through the water. Its large, webbed hind feet enable it to propel itself forward or backwards with ease.

Illus. 17.   The tropical clawed frog, Xenopus tropicalis, has a dorso-ventrally flattened body, typical of aquatic frogs. Because of their position, its eyes can protrude above the surface of the water.

Illus. 18. *The tropical clawed frog is a totally aquatic amphibian. This pair was photographed while still in the water.*

The tropical clawed frog has another interesting adaptation for life underwater. The marks on each side of its body that look like stitches are actually *lateral line organs* (Illus. 17). These organs allow the tropical clawed frog to sense movement in the water.

The tropical clawed frog is not easily identified. You will have to look carefully to distinguish it from other species. It does not have webbed fingers on its front feet as do the dwarf clawed frogs of the genus *Hymenochirus*. It has a tubercle on the inside edge of the underside of its hind foot. It has smaller eyes than other members of the genus *Xenopus*, and it has a tentacle under each eye. It reaches an average snout-to-vent length of under 3 inches. With careful examination, all of these traits will enable you to identify this species (Illus. 18).

The body color of the tropical clawed frog is usually some shade of olive, with faint black or dark green marks. The underside of the body is some shade of lighter green or grey, and there is usually some flecking on the chin. Like other clawed frogs, it has three dark horny structures on the ends of its rear toes that look like claws.

A 10-gallon aquarium makes a suitable home for a pair of tropical clawed frogs. This can be half-filled with dechlorinated water and provided with a substrate of aquarium gravel.

Your local aquarium store can help you to select an appropriate submersible canister filter or an off-the-back filter to keep the water clean. See the section on the Amboina box turtle for a further discussion of these filters.

Although these frogs do come to the surface to gulp air, provide them with a small air pump and an air stone. This helps to oxygenate the water for the tropical clawed frog, since it does perform some respiration through its skin. Once again, your local aquarium store can help you to select the appropriate equipment.

Cover any large openings in the top of this amphibian's tank with a screen, since these frogs can propel themselves out of the water. If they end up on the floor, they will die rather quickly. A lower water level also helps to prevent this kind of accident.

Provide tropical clawed frogs with some sort of platform which protrudes from the water in their aquarium. This platform allows them to hang onto something if they come to the surface to breathe.

The water temperature should be maintained between 75° and 80°F. A standard fully submersible aquarium heater will allow you to do this. Use a thermometer to monitor the temperature. You should feed the tropical clawed frog a varied diet which can include live, frozen, or freeze-dried tubifex worms, brine shrimp, commercial stick foods or pellet foods (types that sink to the bottom are best), and smaller "feeder" fish.

If you use "feeder" fish, remember that some fish will learn to avoid the frogs and they will then survive in the tank. You can keep peaceful community fish in the tank along with the tropical clawed frogs, but you should build small caves of rock or plastic at the bottom of the tank so that the frogs have a secure place to hide. This will prevent the fish from nibbling at the frogs' skin.

The tropical clawed frog will provide you with amusement as it swims actively through its tank, comically shovelling food into its mouth. Although these frogs are not always easy to find and to identify, they make excellent aquarium pets.

Marine toad

Although scientists do not make a distinction between frogs and toads, the members of the genus *Bufo* are commonly known as the "true toads." These amphibians differ from frogs in that they usually have a drier, more warty skin, they lay their eggs in strings rather than masses, and they tend to be able to venture further from bodies of water for longer periods of time.

The marine toad, *Bufo marinus*, is a typical member of the genus *Bufo*. This animal is not strikingly colored, but what does set it apart is its size. Adult marine toads can reach a snout-to-vent length of over 9 inches (Illus. 19).

This toad's native habitat ranges from southern Texas through the Amazon Basin of South America. Unfortunately, it has been introduced to a number of areas to control agricultural pests. It adapted so well to those environments that it too became a pest.

Many areas have laws forbidding the possession of *Bufo marinus* as well as some other species of giant toads. Check all laws in your area concerning these animals before attempting to acquire one.

Giant toads tend to have narrower tongues than do frogs of the same size. These toads are better suited to

Illus. 19.   The giant marine toad, Bufo marinus. *Because it has become a pest in some of the areas of the world where it was introduced, it is illegal in some places.*

eating large insects (as opposed to rodents), so they must be offered relatively large quantities of insects. In fact, some specimens will eat several hundred large crickets in the course of one week. Buy crickets in bulk from a bait supplier. Many of these suppliers will even ship crickets to you in bulk.

Other large insects (dusted with vitamin powder) will enhance this toad's diet. Several newborn mice can be used as a nutritional supplement on a monthly basis.

Due to this animal's large size, an aquarium of at least 15 gallons (with a screen cover) is recommended, and an even larger tank is desirable.

Newspaper is an ideal aquarium substrate for this toad. Whenever this toad enters and leaves its water dish (which also needs to be roomy and heavy), it invariably drags large amounts of gravel or dirt substrate into the water and large amounts of water onto the floor of its tank. Newspaper works well as a substrate because it dries quickly and can be changed easily.

If the room in which the marine toad is being housed does not have a stable temperature, provide supplemental heat by using 25-watt red aquarium bulbs. A temperature between 70° and 80°F is ideal to maintain this animal for years. Use a thermometer to check this.

Like many other *Bufonids*, the marine toad relies on its parotid glands (located behind each eye) to provide it with a means of defense. When this toad is excited, these glands secrete a rather powerful toxin. This toxin has been known to kill large dogs that have mouthed these toads. Keep family pets away from this toad, and wash your hands after handling the toad.

Don't keep this toad's tank in the bedroom; since this toad is a nocturnal creature, it can be quite active and noisy at night.

Green toad

This species, *Bufo debilis*, presents a striking contrast to *Bufo marinus*. It is usually only from 1½ to 2 inches in snout-to-vent length (Illus. 20). Its color is a bright green to yellow with small black dots or bars.

This tiny toad inhabits the drier regions of the central United States and Mexico. It usually breeds in pools of water created by the summer rains that occur in those regions.

A small aquarium with a screen cover and a substrate of coarse gravel or newspaper will make a suitable environment for this toad. A water dish should be provided with a stone ramp for climbing into the dish and a stone inside the dish to assist the toad's departure.

Due to the tiny size of this animal, it must be supplied with very small crickets (or other invertebrates) powdered with vitamins. Use common sense to determine the amount you feed it. Due to its small size, the green toad should not be left for more than a few days without food.

Because they are relatively shy animals, these toads should be supplied with hiding places. Pieces of bark, or a small inverted plastic dish with a hole in its side will serve the purpose well.

Keep the temperature at a stable 65° to 75°F. If you take good care of the green toad, it will provide you with amusement as it takes tiny leaps to move from one place to another.

The green toad is not commonly found in pet stores, but if you manage to find one, it makes a great pet for those who seek the unique.

Illus. 20. Bufo debilis, *the green toad. This tiny amphibian rarely exceeds 2 inches in length.*

Asian tree toad

This toad, *Pedostibes hosei*, is one of the strangest members of the genus *Bufo*. It has long, thin limbs and large adhesive disks on the ends of its fingers and toes to aid it in climbing (Illus. 21).

The toad's color is usually some shade of brown or black with yellow spots. Females can attain a length of 4 inches, while males usually don't exceed 3 inches from snout to vent.

The Asian tree toad is native to Southeast Asia, where it roams on the ground and through low-lying vegetation in search of insects. In the wild, it feeds mainly on

Illus. 21. The Asian tree toad, Pedostibes hosei, *is one of the strangest members of the Bufonid family. The adhesive pads on the ends of its toes aid it in climbing.*

large tree ants, but a supply of crickets or other large insects (dusted with vitamin powder) will meet its dietary requirements.

Provide a roomy aquarium. Because this amphibian's native habitat is in rain forests, near water, the screen top should be partially covered to retain humidity.

Substrates of potting soil and sand, coarse gravel, or newspaper are all acceptable, as long as sufficient humidity is maintained. Use a sizeable, heavy water dish.

The temperature should be kept between 75° and 80°F. Drops in temperature should be avoided, and if you must provide heat with red aquarium bulbs, exercise caution to prevent the tank from becoming too dry.

This animal is rarely available, but if you want a unique toad, the Asian tree toad is a great choice.

5
LIZARDS

Solomon Island giant skink

This large reptile reaches a snout-to-vent length of over 1 foot. It is also known as the zebra lizard which, like its scientific name, *Corucia zebrata*, refers to the faint bands found on the body of some specimens. Another name for this animal is the prehensile-tailed skink. Its thick, muscular tail is used as a climbing aid in this creature's arboreal habitat.

This, the largest member of the skink family, lives in the tropical forests on a number of the Solomon Islands. It uses its prominent claws and prehensile tail to help it reach vegetation among the branches of the trees.

The Solomon Island giant skink has a bulky body. Its smooth scales give its skin an almost plastic feel. It is usually some shade of dull green with brown or black markings.

In captivity, this skink should be misted on occasion (2 to 3 times a week is sufficient) to help it shed its skin. Moderate to high humidity is recommended.

A large aquarium with a screen cover is a suitable container. A taller tank is better than a long, flat aquarium, due to the skink's natural desire to climb.

A substrate of potting soil, bark mulch, indoor-outdoor carpeting or even newspaper is acceptable. Add a hiding place on the ground (for the skink's security), and some large, sturdy branches for climbing, and the lizard will have its basic home furnishings.

A large, heavy water bowl should be used to help maintain a high humidity, and to prevent the skink from tipping the bowl and spilling its contents.

Fresh vegetables such as romaine, escarole and endive lettuces, grated squash and a variety of other chopped vegetables and fruits form the skink's basic diet. Avoid oranges or other citrus fruits. Their high acid content can cause this animal to become ill.

Offer food 4 to 5 days each week, and dust the food with vitamin powder, as well as with a calcium supplement. To provide protein, sprinkle some finely chopped hard-boiled egg on top of the food once every other week.

Focus a 75-watt red spotlight onto one end of the tank to provide an area for the Solomon Island giant skink to raise its body temperature. This also allows the lizard to retreat to the other end of its home to cool itself. Maintain a general temperature between 70° and 80°F.

This skink can use its large jaws to bite aggressively, but most captive specimens become very tame and do not seem to mind being handled as long as they have a sturdy support (Illus. 22).

If you plan to handle this animal, trim its nails, since the nails are quite sharp and can inflict scratches as the lizard climbs. To prevent bleeding, clip only the tip of each nail. If you are unsure about this procedure, a veterinarian can teach you the proper method.

These animals can be expensive if found on the pet trade. To protect what may be an expensive investment, take stool samples of a newly acquired specimen to a veterinarian. Have the animal checked and treated for parasites if necessary. This can prevent expensive and distressing consequences.

Because of the Solomon Island giant skink's interesting appearance and adaptability to captivity, it qualifies as a great unique pet.

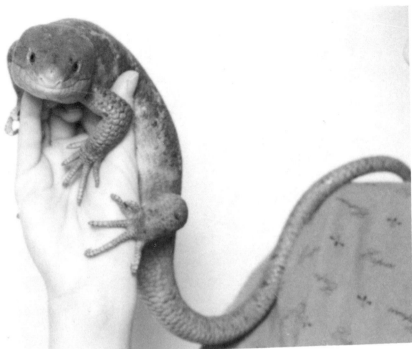

Illus. 22.   The Solomon Island giant skink is one lizard that does not seem to mind being handled, as long as it is provided with good support.

Eastern glass lizard

The Eastern glass lizard, *Ophisaurus ventralis*, is probably the reptile that is least appreciated for its uniqueness. Most people who first see these lizards in a zoo or in a pet store don't seem to grasp the fact that what they are seeing is actually a lizard which has lost its legs through evolution, and not just a strange-looking snake. In spite of this reaction, the eastern glass lizard will still make an excellent unusual pet.

These lizards belong to the family Anguinidae or "lateral-fold" lizards. They have this name because of the long "crease" that runs down each side of the body (Illus. 23). This "crease" gives them a large eating capacity since it allows them to "expand" slightly as they ingest more food.

Illus. 23.   The eastern glass lizard, Ophisaurus ventralis, *has small external ear openings, one characteristic that differentiates it from snakes.*

This lizard's movable eyelids and external ear openings indicate that it is indeed not a snake.

Ophisaurus ventralis inhabits the southeastern United States, and frequents areas where thick undergrowth would catch the legs of a typical lizard as it attempted to escape from predators or to catch prey. The advantage of being legless in such an environment is obvious.

Another unusual trait of this lizard is its tail. If a predator grasps the lizard's tail, the glass lizard will twist its body and lose its tail. The detached tail will wriggle violently, capture the predator's attention, and allow the more vital "front half" of the lizard to crawl away to safety.

This lizard can reach an overall length of over 3 feet, while the snout-to-vent length rarely exceeds 12 inches. This indicates just how long the detachable tail can be.

As a subject of folklore, it was once a commonly held belief that when the eastern glass lizard lost its tail, the two parts met later to reattach. This is a tribute to the length of time that the tail continues to wriggle after it becomes detached.

Remember the detachable tail when attempting to handle the eastern glass lizard. As its name and brittle feel suggest, it can lose its tail easily while it's in your hand. Be gentle and cautious when holding this creature. Give it enough slack to permit it to rotate freely in your hands. Although it can "regrow" its tail, the new one is never as nice-looking as the original.

Use a substrate of bark mulch, potting soil, gravel or indoor-outdoor carpeting in this lizard's vivarium.

Provide these animals with security by adding a hiding place in their screen-covered aquarium.

Use a medium-sized water dish with a stone in it. The stone will prevent food animals from drowning. A variety of foods will supply the proper nutrition. Vitamin-powdered crickets and other invertebrates such as waxworms (*Galleria melonella*) make a good basic diet. Add an occasional earthworm or a newborn mouse to the lizard's diet as a nutritional supplement.

One dozen large crickets introduced on a weekly basis will usually last a moderate- to large-sized glass lizard for one week. Offer other food when it seems appropriate.

This lizard is quite attractive, having irregular black stripes and bars on a green background. Leave a full-spectrum fluorescent light on for 10 to 12 hours each day to help show off the lizard's colors. The light also provides the basking eastern glass lizard with supplemental vitamin D_3.

Maintain the temperature for this animal between 70° and 80°F. These animals do not seem to need electric "heat stones," so heat can be provided by 25-watt red aquarium bulbs.

The eastern glass lizard is reported to live over 15 years in captivity; it is relatively inexpensive, and it makes a hardy terrarium pet. Keep these animals alone or in pairs, since in their wild state, they're known to eat other lizards.

If kept under similar conditions, three other glass lizards also make great reptile pets. Two are from the United States. The first, the slender glass lizard (*Ophisaurus attenuatus*) is browner-colored; the second, the island glass lizard (*Ophisaurus compressus*), is smaller, measuring about 2 feet. The third lizard, from Europe and southwestern Asia, has two common names: the giant legless lizard, or the sheltopusik. This lizard (*ophisaurus apodus*) measures about 4 feet (Illus. 24).

Illus. 24.   The external ear openings can be clearly seen on this giant legless lizard, Ophisaurus apodus.

Rankin's dragon

Don't let the common name fool you. This lizard, *Amphibolurus rankini*, native to Australia, attains an overall length of about 7 inches. Its snout-to-vent length averages 4 inches, so it is hardly a dragon. Its white and beige coloration, robust body, and the spikes running down its sides, give this lizard (from the family Agamidae) a striking resemblance to the horned lizards (*Phrynosoma* species) of western North America (Illus. 25).

Unlike the horned lizards, Rankin's dragons do very well in captivity if fed a diet of crickets, waxworms (*Galleria melonella*), and leafy greens (like romaine) dusted with vitamin powder.

This lizard needs heat to remain healthy. Provide this reptile with an area of its cage that reaches a temperature of over 90°F. Use a white "spot" bulb of 40 to 75 watts shining through the glass into one corner of the tank. Turn off the spot bulb at night; a slight temperature reduction in the evening is good for this lizard—as long as the general temperature does not fall below 75°F. An electric "heat-stone" alone will not provide enough heat for this lizard to digest its food. Give the lizard an area of the terrarium to escape excess heat and to cool itself.

This docile lizard does not attempt to bite, but it will attack other lizards, even if they are much larger than it. Keep this animal with similar-sized members of its own species. Don't keep more than one male in a tank—males may fight for territorial dominance.

This particular species is not recognized in many books on Australian reptiles, so there is some question as to its actual scientific name. *Amphibolurus rankini* is the name that is currently being used in the United States by the breeders of this lizard.

Due to Australia's ban on the export of native species, most of the specimens you may encounter will probably be captive-bred. Because of this ban, you may acquire a

Illus. 25. Rankin's dragon, Amphibolurus rankini, *is a member of the Agamid family. It is native to Australia.*

hatchling, and not an adult. It is vital that captive-bred hatchlings of this species be offered a dozen small crickets and some green vegetables at each feeding, at least *five to six days each week.* Every-other-day feedings are not sufficient for specimens under 6 months of age. The hatchlings will die of malnutrition if you don't follow the heavier feeding schedule.

The reason for the extraordinary amount of food which the young lizard consumes is that the animal will almost double in size during its first 4 months of life. This growth requires a great deal of energy. The Rankin's dragon will use its extensible tongue to catch and eat as many insects as it can in order to satisfy its voracious appetite.

Since it is a desert animal, expose the Rankin's dragon to 10 to 12 hours of full-spectrum fluorescent lighting each day. This provides the lizard with its necessary dose of vitamin D_3.

The Rankin's dragon can be maintained in a glass aquarium with a screen cover and with a substrate of either sterile sand, aquarium gravel or indoor-outdoor carpeting, along with a small water dish for an occasional drink.

If the Rankin's dragon is provided with the proper care, it will live for years in captivity and it will make a superb unusual pet.

Chinese crocodile lizard

The Chinese crocodile lizard, *Shinisaurus crocodilurus*, is certainly one of the most unusual lizards on earth. It looks like a cross between a lizard and a crocodile (Illus. 26). The Chinese crocodile lizard was first described in 1928, making it a relatively recent discovery. It has an olive coloration and a laterally compressed tail that serves the animal well in swimming and diving.

This reptile inhabits the Dayao Shan mountain range in the eastern part of the Chinese province of Guangxi. It lives there near the pools formed by slow-moving streams.

The Chinese crocodile lizard is diurnal and becomes active in the morning when it begins to hunt for insects (both terrestrial and aquatic), as well as for tadpoles and for the fish that make up its diet.

This lizard likes to both bask and sleep on branches which hang over the water. At the first threat of danger, the Chinese crocodile lizard will dive into the water and

Illus. 26.   The Chinese crocodile lizard, Shinisaurus crocodilurus, *is a reptile that is superbly adapted for an am-
phibious existence.*

swim to the bottom, where it will be submerged for up
to 30 minutes before returning to the surface.

This animal, a member of the lizard family Xenosau-
ridae, is unusual because its preferred body temperature, between 70° and 72°F, is somewhat lower than
that of most other lizards.

The Chinese crocodile lizard reaches an average
snout-to-vent length of about 10 inches, and an overall
length of 16 inches. Because of the lizard's size, use an
aquarium of at least 15 gallons (with a screen cover) for
captive maintenance. A larger aquarium is even better.

There are two ways that you can keep the Chinese
crocodile lizard successfully. The first is to use a substrate of newspaper or coarse aquarium gravel and to
provide a large water dish with stone ramps on the inside and the outside of the dish for access to the water.
Although this type of setup does not allow you to observe a great deal of the lizard's swimming and diving

behavior, due to the shallowness of the water dish, it does offer the advantage of easy cleaning and water changing which both help to ensure the health of your animal.

The second way to keep this reptile successfully is in an aquatic terrarium. Divide a 20-gallon tank vertically with a 3½-4-inch high piece of glass, held in place with aquarium silicon adhesive. Fill half the tank with water to a depth of 2 or 3 inches, and then fill the other half with gravel. Build a ramp from one side of the tank to the other. Or, simply fill the tank with water, and then build one or two rock islands. Both methods call for a filter system. Consider the disadvantages of more difficult cleaning and maintenance. These two tank setups allow you to observe the animal's swimming and diving behavior.

If you choose the second type of setup, your local aquarium store can help you choose an appropriate submersible canister filter to keep the water clean. The store can also help you with the construction of the aquatic terrarium.

No matter how you maintain the Chinese crocodile lizard, there must be an area where the lizard can get completely dry. If you do not provide such an area, the lizard will develop skin lesions, and it will eventually die.

Maintain the air temperature of the tank between 68° and 75°F, and be sure to use a thermometer to monitor the environment accurately.

You should use a full-spectrum fluorescent bulb between 10 and 12 hours each day to provide the Chinese crocodile lizard with enough vitamin D_3 to maintain its health.

A diet of crickets (dusted with vitamin powder), feeder goldfish and an occasional tadpole or newborn mouse will provide the proper nutrition. You can introduce a dozen large crickets on a weekly basis for each lizard you have, and then offer the other food items as you see fit.

Chinese crocodile lizards usually become quite tame in captivity and do not seem to mind being handled occasionally. They make great pets, and are quite fascinating to observe as they swim and catch their prey.

Because they come from such a limited area, Chinese crocodile lizards are not always available, but if you are able to acquire one, you will have a unique reptile.

Blue-tongued skink

This lizard, *Tiliqua scincoides*, may be one of the best lizard pets to be found on the market. It eats a variety of foods and it also becomes extremely tame, with an easy-going disposition.

This member of the skink family is native to Australia, where it ranges from the extreme north through the east and southeast.

Two subspecies of this lizard are recognized. The first, *Tiliqua scincoides intermedia*, is found in the north of Australia, and is distinguished by a lighter orange/brown dorsal color. The second subspecies, *Tiliqua scincoides scincoides*, is found to the east and southeast of Australia, and can usually be distinguished by a darker dorsal color and a more pronounced streak on each side of its head, continuing back from its eyes.

Many consider the first subspecies to be more attractive in coloration and therefore more desirable as a pet. If you are a true lizard enthusiast, you will see that both subspecies are uniquely marked. Both make equally good pets.

This lizard gets its name from its brightly colored tongue. When attacked or threatened, the blue-tongued skink extends and unfolds its tongue, which is almost as large as its head. The tongue's bright blue color serves to frighten and to discourage attackers or predators.

The general appearance of this animal also makes it unique (Illus. 27). It attains an average snout-to-vent length of 12 inches, with an additional 8 inches of tail. The body of the blue-tongued skink is cylindrical and quite robust, while the legs are relatively short. The back of the reptile is marked with dark cross-bands.

Because of the large size of its body and the shortness of its legs, the blue-tongued skink does not raise itself to walk. Instead, it slides on its belly as it moves. Surprisingly, this lizard is capable of fast movement over short distances, allowing it to catch and to eat live prey.

The blue-tongued skink defends itself by flattening its body and hissing loudly. Some scientists believe that this is an attempt to mimic the poisonous death adder snake, *Acanthophis antarcticus*, which does share some of the skink's range. Both animals can have similar coloration and size, and the blue-tongued skink's small legs could cause an animal to mistake it for a

Illus. 27. Although the blue-tongued skink, Tiliqua scincoides, *can be expensive, it makes one of the best lizard pets.*

death adder. In fact, when man meets skink, this apparent mimicry can work to the lizard's disadvantage. In Australia, many lizards have been killed when they were mistaken for the poisonous death adder.

In fact, the blue-tongued skink is not dangerous at all, and it will quickly become tame in captivity. It will learn to take food from your hand, and it will even come to the front of its tank when you approach.

The blue-tongued skink inhabits forest, woodland, grassland and semi-arid areas in Australia, so a substrate of bark, aquarium gravel, newspaper or indoor-outdoor carpeting will suit it nicely.

Because of the skink's large size, a 40-gallon aquarium (with a screen cover) is recommended. Include several hiding places, a large rock for climbing, and a medium-sized water dish, and you will have a good basic environment.

The general temperature of the terrarium should be maintained between 75° and 85°F. Use a 75-watt red spot bulb focused onto one end of the tank to maintain this temperature and to provide a warm area for basking.

Leave a full-spectrum fluorescent light on over the tank for 10 to 12 hours each day. The lizard's food should be powdered with vitamins to ensure the proper levels of calcium and vitamin D_3.

Feed the blue-tongued skink a variety of foods, including crickets, young mice, romaine lettuce, snails, cottage cheese, dog food, grated squash, strawberries, broccoli and melon. To supply proper nutrition, vary this lizard's diet as much as possible. Fresh food should be offered 3 to 4 times each week.

The blue-tongued skink's nails can become quite sharp, and if you trim them, do it very carefully. The nails are relatively short, and it is easy to hit a vein and cause unnecessary bleeding. The best advice is to avoid trimming its nails and to handle the lizard carefully. If you must trim the nails, have a veterinarian who specializes in reptiles teach you the proper method.

Although this lizard's tail is prehensile, it can be lost. Never grab this lizard by the tail; hold the lizard firmly when handling to prevent falls and unnecessary tail breakage.

Because Australia does not allow the export of its native animals, the captive-bred blue-tongued skink can be quite expensive. There are a number of breeders of this animal, and the blue-tongued skink's adaptability to captivity and its gentle disposition make it one of the best reptile pets, despite its high price.

False monitor

This lizard, *Callopistes flavipunctatus*, is not really a monitor at all; it is actually a member of the Teiidae family. It is also known as a monitor tegu because of its resemblance in appearance to tegu (teju) lizards. The false monitor is native to Ecuador and Peru, where it inhabits the rocky deserts near the coasts.

The false monitor averages an overall length of over 3 feet. Its body is usually marked with shades of brown and beige with bright yellowish dots. Like other Teiidae, it has a long, slender, cylindrical body and tail, with long toes and prominent claws (Illus. 28 and 29).

This reptile can run quite fast. It takes advantage of its speed to hunt other lizards, its primary prey in the wild. The false monitor in its wild state will also eat small mammals and insects.

The false monitor is diurnal. It retreats to burrows or small caves at night and emerges in the morning to bask and hunt for food as the temperature rises.

One interesting aspect of the false monitor's native environment is the great temperature fluctuation. Because the deserts it inhabits are near the coast, there is a significant temperature drop at night. The false monitor thus has to deal with a wider temperature range than do many other reptiles. The false monitor takes full advantage of the daytime heat when the temperature is ideal for hunting and for other activities; it rests and conserves energy at night when the temperature is lower and not at the optimum level.

The false monitor is known as a *foraging predator*—it actively searches for prey. An *ambush predator* waits for prey to come to it.

Because of its large size, house the false monitor in an aquarium of at least 40 gallons—a larger tank is preferable. A screen cover serves as a suitable top for the tank. The substrate can be either newspaper, indoor-outdoor carpeting or bark chunks.

Provide a medium-sized heavy water bowl. A sizeable hiding place of wood or cardboard will give the false monitor the sense of security that it requires.

Although the false monitor eats lizards in the wild, rodents will serve as a proper diet in captivity. A weekly feeding of one to several mice (depending upon the false monitor's size) will supply the lizard with proper nutrition. Some specimens can be taught to take dead mice;

Illus. 28. The false monitor, Callopistes flavipunctatus, *from Ecuador and Peru is rarely imported.*

Illus. 29.   A close-up of the false monitor's snout.

this will allow you to keep frozen mice on hand and to thaw them as needed.

Temperature is a key element in maintaining this animal in captivity. Because of the temperature fluctuations in its natural environment, provide a temperature drop at night. The general temperature of the false monitor's tank should be maintained between 75° and 85°F in the daytime. Focus a 75-watt spot bulb on one end of the tank to maintain the temperature. The spot bulb does not have to be red, it can be placed on a timer so that it turns off after 10 to 12 hours. This allows the temperature to drop at night. Allow the nighttime temperature to fall to between 60° and 70°F. Use a full-spectrum fluorescent bulb over the false monitor's tank between 10 to 12 hours each day. This supplies the lizard with the proper levels of vitamin D_3.

The false monitor is a mild-tempered animal. It makes an excellent pet and it generally does not attempt to bite when in captivity.

This lizard exhibits an interesting type of behavior when it feels threatened. It "plays dead" and will become limp in your hand. Once it feels that the danger is past, it will then resume its normal activity.

Because of its remote habitat, the false monitor is not often imported. It commands a high price if and when it does become available. Before you buy this lizard, check it thoroughly. Take a stool sample to a veterinarian just to be on the safe side.

The false monitor is quite an attractive lizard, with a relatively gentle disposition. If you acquire one, you will have a great unusual pet.

6
SNAKES

Trans-Pecos rat snake

The Trans-Pecos rat snake, *Elaphe subocularis*, is one of the most attractive of all North American snakes. Its body is a tan color, with varying shades of brown that form "H"-shaped markings down the snake's back. Because of these markings, it is also called the "H" Snake (Illus. 30 and 31).

This snake's large eyes have distinctively round pupils. This is particularly unusual since the Trans-Pecos rat snake is primarily nocturnal, and most nocturnal reptiles have vertical pupils.

Aside from its outward appearance, an advantage to owning this snake is that it never reaches a size which makes it difficult to manage. Many people select baby boa constrictors or pythons without realizing that some species get *very* large and can become irritable as they grow older. The Trans-Pecos rat snake reaches an adult length of only 5½ feet and will remain quite tame.

This snake inhabits the rocky areas of the Chihuahuan desert of south-central North America. The vivarium for this animal should be kept relatively dry. It only needs a medium-sized water dish.

An aquarium of 15 gallons or more (with a screen cover) makes a suitable home. Use a substrate of sand/soil or sand/peat moss mixtures or pine shavings.

Illus. 30.   Elaphe subocularis, *the Trans-Pecos rat snake. The markings on its back resulted in its nickname, the "H" snake.*

Do not use cedar in any form or any prepared materials (such as cat litter) for substrate, since these contain oils and chemicals which can irritate the snake's mucous membranes.

Provide a hiding place as a retreat for these snakes, and include a sturdy branch or two for climbing.

Maintain the temperature of the Trans-Pecos rat snake's environment between 75° and 85°F. This can be achieved easily by using red aquarium bulbs.

A common misconception about rodent-eating snakes (such as this one) is that they will only eat live food. This is not true, so you can feed your snake pre-killed rodents to prevent a live mouse from injuring the snake. Mice sometimes attempt to give one last bite as a snake constricts them. If you *must* feed a snake a live rodent, never leave the cage unattended. If the snake does not

Illus. 31. This photograph clearly shows the "H" pattern on this snake's back.

have a feeding response, and the mouse is left in the vivarium, the mouse may turn and damage the snake's eyes or face and possibly even kill it.

Some snakes require "training" before they will accept pre-killed mice. Move the mouse in front of the snake, using tongs or some other device. Sometimes leaving a pre-killed rodent in the snake's cage overnight will encourage feeding, as well.

Once a snake is trained in this way, feeding will become easier. Many pet stores sell frozen mice—the snake keeper will only have to thaw and feed.

For hatchling snakes feeding on newborn mice, feedings can be conducted on a weekly basis. Full-grown adults may require 2 adult mice per week to satisfy their appetites.

When these snakes are about to shed their skin, they may occasionally skip a weekly feeding. You'll know that your snake is ready to shed its skin when its eyes have a cloudy appearance. This is no cause for alarm.

A snake should never be overfed to the point where its body bulges quite noticeably. If this happens, the snake will be unable to digest its meal and may actually regurgitate it.

Feed the snake somewhere other than in its home aquarium. Transfer the snake into a large paper bag. This prevents the snake from associating food with a hand entering its cage. Such an association could lead to an accidental bite. It is especially important to feed in the paper bag if the snake is not handled on a regular basis. This also forces the owner to hold the snake at least once each week.

One key point to keeping the Trans-Pecos rat snake successfully is cleanliness. Be sure to remove the snake's waste promptly and to change the substrate on a regular basis. This is a good way to assure the snake's health.

It should be noted that the state of Texas protects the Trans-Pecos rat snake, but there are a number of people who captive-breed this species. You can buy a captive-born specimen without putting any undue strain on the wild population.

If you want a unique snake that will become very tame in captivity, the Trans-Pecos rat snake is an excellent choice.

Sunbeam snake

This interesting snake, *Xenopeltis unicolor*, is native to southern India and Southeast Asia. Its common name comes from the fact that is has iridescent scales. When exposed to light, these scales reflect a rainbow of colors.

The color of this snake's upper body is usually some shade of brown or black; it has a whitish underside. Adult specimens average from 2 to 3 feet in length, with a rather long, narrow head. When excited, it will vibrate its pointed tail.

The sunbeam snake is a member of the family Xenopeltidae and is a *fossorial* reptile, meaning that it spends most of its time burrowing in soft soil near rivers, streams and lakes, where it hunts for frogs, small snakes and rodents (Illus. 32).

A glass aquarium with a screen cover makes an ideal container for the sunbeam snake. Maintain a temperature between 70° and 80°F. Red aquarium bulbs are a great way to heat the tank, if needed.

A substrate of potting soil, sphagnum moss or bark mulch is essential because of this animal's high humidity requirements. The substrate should have a portion which is kept somewhat moist, and a medium-sized bowl of water is also required.

The sunbeam snake spends most of its time under the soil. Place a glass plate on top of about ½ inch of the soil, to let you see the snake at all times. The snake will still have the security it needs.

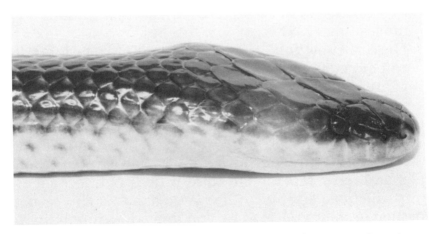

Illus. 32. The sunbeam snake, Xenopeltis unicolor, *is a burrowing snake.*

Feedings should be conducted on a weekly basis; one young mouse per week will provide all the necessary nutrition. Most sunbeam snakes cannot digest a full-grown mouse, so that is why smaller food is suggested. Frogs and lizards can be offered for variety, but they are not necessary. Freshly killed and even thawed frozen mice will be taken by most specimens, and this can be quite convenient for the keeper. The sunbeam snake will eat other snakes, so keep it in a tank by itself.

The sunbeam snake is often overlooked as a pet. This is unfortunate, since it is reportedly abundant in its native habitat. It doesn't bite under most circumstances, it eats heartily in captivity, and it lives for years if given the correct care. Considering all these facts, the sunbeam snake makes a great pet for either the beginning or the advanced snake keeper who desires the unusual.

Carpet python

For those of you who absolutely *must* have a python, the carpet python, *Morelia spilota variegata*, has many advantages (Illus. 33).

The carpet python can reach a length of 10 feet, but it does not get as bulky as many other boas and pythons that are available through the pet trade; therefore, this snake is easier to handle than the other boas and pythons, and unlike many large snakes which require chickens or larger prey, the carpet python can be fed a diet of rats throughout its adult life. The carpet python has an even temperament, and it does not become as unpredictable as some other large snakes. The carpet python becomes quite tame and it is a pleasure to handle. It breeds readily in captivity, and there are a number of breeders of this animal, so although it is unusual,

Illus. 33.   The vertical pupil of this carpet python, Morelia spilota variegata, *is characteristic of many snakes that become active at night.*

the carpet python is not as expensive as some other pythons.

This snake is native to Australia and has varying patterns of beige and brown background with black or dark brown splotches on that background.

Like many other pythons, the carpet python has a row of heat-sensitive labial pits on each side of its face. These pits help the snake to locate its prey by detecting the prey's body heat. The pits are capable of sensing changes in temperature as slight as a fraction of a degree and serve the snake well in its quest for food.

The carpet python can be kept in a 40-gallon aquarium with a sturdy screen cover, a substrate of pine shaving or bark mulch, and a sturdy branch for climbing.

Maintain the temperature between 75° and 85°F by using red light bulbs. Use a large, heavy water dish to prevent spillage. Mist the snake's body occasionally to assist in skin shedding.

Weekly feedings of pre-killed mice or rats, depending on the size of the snake, will ensure the reptile's health. Do not overfeed it to the point where its body bulges considerably. As a guideline, never feed this snake anything bigger than 1½ times the size of its head.

This is one snake that definitely should be fed in a location other than its vivarium, since carpet pythons are particularly ravenous eaters when a feeding response is induced. In order to satisfy their appetites, they will strike at anything that moves.

Another subspecies of the carpet python, the diamond python (*Morelia spilota spilota*), can be maintained in the same manner and reaches a similar length. The diamond python is black with yellow dots and diamond-shaped markings on its body. It is also rarer in collections and is considerably more expensive than the carpet python.

If you seek a python as a pet, the carpet python makes a good choice because it eats well, it lives for years and it remains tame. It is unusual since it is not found as commonly in captivity as are many other large constrictors.

Sand Boa

Sand boas, members of the family Boidae, inhabit areas of Africa and Asia. They could be more accurately called "earth boas," since they are not limited to areas of sandy soil. They can sometimes be found in places with rich vegetation.

Sand boas are members of the genus *Eryx*. The 10 species in this genus range in size from an average of 15 inches, the size of Mueller's sand boa of Africa (*Eryx muelleri*), to almost 30 inches, the size of the brown sand boa of Asia *(Eryx johnii)*.

To escape the heat of the day and to avoid predators, sand boas generally stay just below the surface of the sand. This also gives them an excellent vantage point from which to ambush their prey.

These snakes all possess special characteristics that permit them to burrow and to live under the surface of the ground efficiently. One of the most noticeable of these characteristics is its blunt, wedge-shaped head (Illus. 34). The nostrils are located on the sides of the head; the lower jaw fits tightly into the upper jaw, which projects out slightly. Both of these characteristics work in combination to prevent sand from getting into the snake's nose and mouth.

Illus. 34.   The head of a typical sand boa with its wedge shape and small eyes. This is the Egyptian sand boa, Eryx colubrinus colubrinus.

Sand boas generally spend their days just under the surface of the soil. They emerge at times from dusk until dawn to look for a mate or to actively hunt prey. During the day, sand boas are known also to take refuge under rocks or in rodent burrows.

These snakes have cylindrical bodies and small eyes. In spite of their small eyes, these animals seem to have vision that is comparable to that of most other snakes.

In the wild, sand boas consume a variety of prey, depending upon the species of sand boa. Food items can include lizards, insects, other snakes, small mammals and birds.

Sand boas will sometimes actively seek prey at night. Day or night, the sand boas will dart out from the sand when they sense an animal is walking on the sand above them. They will then seize the prey in their mouths and constrict it before swallowing. As with all constricting snakes, they do not crush their prey, but merely hold it tightly until asphyxiation occurs.

Sand boas are usually overlooked for their value as pets. In a properly equipped terrarium, they do quite well, and they will live for years. Many people do not like the fact that sand boas stay buried for most of the day. Place a plate of glass on the surface of the sand (like the setup described for the sunbeam snake) to allow you to view the animal. Allowing them to burrow in the sand without the glass is usually more interesting to watch.

A 10- to 15-gallon glass aquarium (with a screen cover) makes a suitable container. A substrate of several inches of fine sand allows the sand boa to show its extraordinary burrowing abilities. Beach sand should never be used for a substrate due to its salt content (which could be a skin irritant) and possible parasitic contamination. Newspaper (with a hide-box) can also be used as a substrate, although it certainly does not provide a dramatic burrowing spectacle.

An area of the sand should be kept warm. Focus a red spot bulb onto one end of the tank to achieve this. Use a heating pad under the terrarium or use commercial "heat tape" to provide a heated area of the sand. The general temperature of the environment should be kept between 75° and 85°F. A slight reduction of 5°F at night is beneficial.

Place a small water dish (that the snake can not easily tip over) in the tank. You might want to place a plastic container (with a lid and a hole cut in its side) in the terrarium, as well. This container should be half filled with damp sphagnum moss since some sand boas appear to prefer this container to a water dish for re-hydrating themselves. You can experiment with a con-tainer of this type to see if your sand boa will use it.

Feed your sand boa on a weekly basis. The size and number of mice you feed it depends upon the size of the snake. Most sand boas will eat readily if they are kept in an environment with the proper warmth. If yours re-fuses to eat, try a "pinkie" (a newborn or hairless mouse) because some small specimens will not eat mice that have fur. Using the plastic container with moss seems to be important to some specimens to maintain their appetites.

It is fascinating to watch these snakes dart out of the sand and catch their prey. If you keep more than one in a tank, separate them when feeding, since they are usu-ally aggressive feeders. This prevents unnecessary inju-ries.

Do not keep different species of sand boas together, since they have been known to attack each other. Mem-bers of the same species usually will not do this.

Two of the most colorful species are the rough-scaled sand boa, *Eryx conicus* and *Eryx colubrinus*. The latter has two common names depending upon the subspe-cies: the Egyptian sand boa, *Eryx colubrinus colubri-nus*, and the Kenyan sand boa, *Eryx colubrinus loveridgei*. These two species and subspecies all average between 1½ to 2 feet in length and make excellent pets.

Add a few decorations of stones or succulent cacti to the sand boa's terrarium to make an attractive display. To find your snake during the day, you will have to sift through the sand in this setup. Most specimens do not seem to mind occasional handling. If you are looking for a unique snake that does well in captivity, consider the sand boas.

Calabar python

Many snakes seem to invite superstition and fear. Some even develop supernatural reputations due to their unusual appearances or behaviors. The Calabar python, *Calabaria reinhardti,* is one such snake.

The Calabar python, native to West Africa, is a snake that burrows through the leaf litter on the floor of the rain forest. Its small, blunt head, cylindrical body, and small eyes are marks of its subterranean lifestyle.

This snake also enters rodent burrows. It constricts its prey either by wrapping around it or by pressing it against the walls of the rodent's burrow.

The color of the Calabar python is generally some shade of brown, with mottled orange or red markings. A unique feature of this snake is its tail, which is usually set apart from the rest of its body by a cream or white border (Illus. 35). This makes the tail look more like a head than the snake's head itself.

The Calabar python uses its tail to deceive predators. It travels with its tail slightly elevated, and it raises its tail and hides its head when attacked. This snake can then survive severe attacks and then escape into the ground.

If the attack continues, the Calabar python will coil itself tightly into a ball with its head positioned in the middle. The tail has a slight spike on the end, and the snake will push this into the attacker, imitating a bite. These ruses have worked over the years to ensure the survival of this unusual snake.

Illus. 35.   This is the tail, not the head of the Calabar python, Calabaria reinhardti. *The fact that the tail is set apart by this white band, deceives attackers into going after the wrong end of the snake.*

Because of its unique behavior, the Calabar python has developed an unusual reputation with many of the natives of West Africa. They believe that this snake has supernatural powers, and they fear it greatly. Some African natives believe that the python is indestructible, since one can strike the "false head" with no effect.

The Calabar python can be maintained in captivity in a 10- to 15-gallon aquarium with a screen cover. Provide a substrate of potting soil and bark mulch to simulate the humidity levels that this snake encounters in its native habitat. Such a substrate also provides an appropriate burrowing medium.

Provide a small, sturdy water dish. Fill a covered plastic container with damp sphagnum moss, and cut a hole in the side for access. In combination, these two devices will meet the Calabar python's water requirements.

Keep the temperature of the animal's tank between 75° and 80°F. Use 25-watt red aquarium bulbs to achieve this.

Offer the snake one small mouse on a weekly basis. If you offer a newborn mouse, you can leave it in the tank overnight. Many Calabar pythons will learn to take pre-killed rodents. This can simplify feedings.

As a general rule, these snakes never bite in defense. This is one reason why they make good pets. Although they are handleable, do not forget that they are quite shy by nature, and too much handling could overly stress the snake. For this reason, do not handle these snakes 24 hours before or after feeding them.

Another good point about the Calabar python is that it averages about 30 inches in length, so it never becomes too large to manage.

If you want what may be the most unusual of all of the pythons, consider the Calabar python as your choice for a unique snake pet.

7
TURTLES

Amboina box turtle

The Amboina box turtle, *Cuora amboinensis* (also known as the Malayan or Philippine box turtle), is frequently overlooked as a prospective pet. Although it is similar in body design to other box turtles, and can close both halves of its shell for protection, the Amboina box turtle is primarily aquatic.

The Amboina box turtle reaches an average shell length of 8 inches. It is dark brown, olive or black with a dark head that has a thin yellow stripe running down each side and across each eye (Illus. 36).

This turtle from Southeast Asia is best maintained in a 40-gallon aquarium that is half filled with water. This tank must have a partially submerged island made of rock, whose top should remain dry. This rock allows the turtle to bask and to dry the top and bottom of its shell.

The island can have a base of brick, cement block or plastic with a "stair-step" design to provide ease of access for the turtle. The top of the island *must* be shale, slate or other dense rock. Clay or cement is too porous and rough and will not dry completely. Such rough materials tend to abrade the underside of the turtle's shell. The island should be sturdy, so that it will not topple over and then pin the turtle underwater.

Maintain the water temperature between 75° and 85°F with the use of a fully submersible aquarium heater. A

Illus. 36.   The Amboina box turtle, Cuora Amboinensis, *can withdraw its limbs completely into its shell for protection.*

25-watt red aquarium bulb can be suspended above the island to keep the air temperature in the same range.

Since these turtles bask to obtain vitamin D_3, a full-spectrum fluorescent bulb should shine into the tank for 8 to 12 hours each day. The fluorescent bulb should be turned off at night to provide the animal with an appropriate day-night cycle.

The water for the Amboina box turtle should *not* be dechlorinated when it is changed, and non-iodized aquarium salt should be added to the fresh water at a dosage of 1 tablespoon for each 5 gallons of water. These two steps will help to prevent the growth of fungus, which can cause potentially fatal shell rot. For an adult turtle, change the water every 4 to 6 weeks.

The water in the tank should be filtered as well, and your local aquarium store can help with the selection of either an off-the-back type filter that has an extension tube to reach the lower water level (Illus. 37), or a power filter that is fully submersible. These submersible filters offer the advantage of minimal splashing; this helps to prevent deposits from forming on the glass (Illus. 38).

Any item you introduce into the tank must be sturdy, since these turtles are quite active and tend to move the objects in their tanks. These turtles are surprisingly good climbers. Don't inadvertently provide them with a means of escape.

The Amboina box turtle can be fed a variety of food, including "feeder" fish, commercial turtle food, and soft fruits (such as plums and strawberries) on an every-other-day basis. This diet will help keep the turtle vigorous and healthy. Many specimens will overcome their

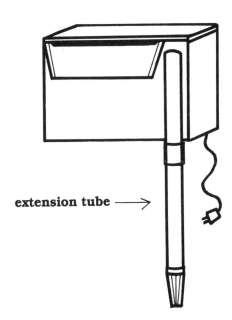

extension tube $\longrightarrow$

Illus. 37. This is an off-the-back filter with an extension tube. This tube is added so that the filter can reach the low water level found in an aquatic turtle tank.

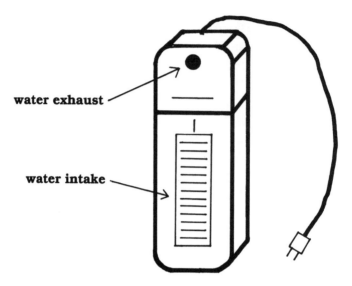

Illus. 38.   This is a fully submersible canister filter. Water is pulled in through the water intake and then through a sponge inside the filter. The water then exits through the water exhaust.

shyness and learn to recognize their keepers and take food from their hands.

If you're looking for an unusual aquatic turtle, the Amboina box turtle in an excellent choice. It thrives in captivity, it does not look like most aquatic turtles, and it is relatively inexpensive.

Pancake tortoise

If you are seeking a terrestrial turtle, the pancake tortoise, *Malacochersus tornieri,* is undoubtedly an unusual choice.

Native to Kenya and Tanzania, this tortoise reaches an average shell length of 6 inches. It comes in a wide range of patterns in cream, brown, beige and black. As its name implies, the pancake tortoise's shell is relatively flat, and the shell is soft and flexible (Illus. 39).

This soft, flat and flexible shell serves this animal well. When it is disturbed, the pancake tortoise does not withdraw into its shell as other tortoises do. Instead, it runs into a rock crevice, where it expands its lungs and then wedges itself between the rocks so that it can not be removed.

This tortoise has slender, flexible limbs which, coupled with the lightness of its shell, allow it to climb the rocky cliffs and outcroppings in its natural habitat. Its flexible limbs also make it easy for this tortoise to right itself should it fall on its back.

A 40-gallon aquarium (with a screen cover) and a substrate of coarse gravel or indoor-outdoor carpeting is recommended. *Never* use sand as a substrate, since tortoises tend to ingest the sand, causing intestinal blockage, which can ultimately result in death.

Keep the tank temperature between 75° and 85°F by using red bulbs. Make one area of the cage warmer by using a spot bulb.

Administer wide-spectrum fluorescent light to the tortoise between 8 and 12 hours each day. A low, secure hiding place made of bark or sturdy rock is necessary for this animal's security.

Feed the pancake tortoise a variety of leafy greens (such as romaine, escarole and endive), and fruits (such as melon and strawberries). For variety, feed it shredded squash or carrots. Serving canned dog food or cat food occasionally will provide extra protein.

Sprinkle the food lightly with reptile or bird vitamins, and give the tortoise its food fresh at least 3 to 4 days each week. On a weekly basis, remove the food for one 24-hour period to allow the animal to fast and to clear its digestive system.

With this much feeding, the tank will have to be cleaned regularly. Remove the waste quickly. The tortoises will be less likely to walk through such waste and make a mess.

Illus. 39. The shape of the pancake tortoise, Malacochersus tornieri, *is a unique design for a land tortoise.*

The pancake tortoise becomes quite tame and will live for many years if maintained properly. Longevity and tameness coupled with its strange appearance and strange habits, make the pancake tortoise a prime choice for one who seeks an unusual pet.

Chinese four-eyed turtle

This turtle, *Sacalia bealei,* gets its common name from the yellowish eye spots that occur on top of its neck, just behind its head. Some specimens have only one pair of these eye spots, while others have two pairs, so this turtle is also referred to as the Chinese six-eyed turtle.

The Chinese four-eyed turtle is native to southeastern China and Southeast Asia. It inhabits mountain streams at both low and high altitudes.

Like most members of the family Emydidae, its claws (on both its front and rear feet) help it to climb over the rocks and logs encountered in its natural habitat.

The Chinese four-eyed turtle's shell is typically a light to dark brown with black streaks and spots on the carapace (the top half), while the plastron (the bottom half) usually has a lighter color (Illus. 40).

This turtle's limbs are usually also brown. Sometimes it has black dots and yellow streaks on the head and neck. Many specimens have a reddish to pinkish color on the inside of the neck and the legs, and even on the plastron. Sometimes the iris of the eyes is yellowish to pinkish in color, making this turtle quite an attractive animal.

The average size of the Chinese four-eyed turtle is about 5 inches (measuring the plastron). Females are typically somewhat larger than the males.

In the wild, the Chinese four-eyed turtle feeds mainly on crayfish, earthworms and fish. Like most turtles, it will eat any other animal matter, dead or alive, that it can get, and if the food is too large, it will use its beak and claws to tear off bite-sized pieces.

In captivity, the Chinese four-eyed turtle can be maintained in much the same manner as the Amboina box turtle.

A 40-gallon aquarium is a good size tank for 1 or 2 specimens, and the special flat "turtle tanks" are usually suitable enclosures for this species.

An island of rock that remains dry while still partially submerged is mandatory so that the turtle can bask and get completely dry. As in the case of the Amboina box turtle, this island can have a base of brick or plastic with "steps" to allow for easy access. Make sure that the island has no rough edges on which the Chinese four-eyed turtle could scrape and abrade its shell, since these turtles seem to be particularly prone to "shell rot."

Illus. 40. The Chinese four-eyed turtle, Sacalia bealei, *is an attractive reptile that usually becomes quite tame in captivity.*

If you acquire a new specimen, check the shell carefully to see if there are any white spots just below the shell's surface. You will be able to see through the very top layer of shell if there are any spots. It isn't unusual for new specimens to have pits in their shells where there once was an active case of "shell rot." This fungal infection is curable in the initial stages. Left untreated, it will eventually be fatal.

Treatment consists of using a blunt instrument to scrape away the white cheesy substance. This white substance is actually a by-product of the infection. It is an indication that the infection is still active.

Once the area has been scraped clean, rinse the area thoroughly and generously apply a solution of either *gentian violet* or *malachite green* to the entire area. Both of these solutions can be found in your local aquarium store. The affected area should be re-treated every 2 to 3 days for a month to ensure that the fungus does not return.

The water in the Chinese four-eyed turtle's tank should *not* be dechlorinated, and non-iodized aquarium salt should be added at a dosage of 1 tablespoon salt for each 5 gallons of water. Both of these steps will help to prevent the formation of shell rot, and will prevent its reappearance on previously-infected specimens.

Specimens infected with shell rot will usually recover fully even if severely infected. The pits in the shell will remain to some degree, but the turtle itself will live for a long time, given the proper treatment.

Maintain the water temperature at a level of around 75°F by using a submersible aquarium heater. The air temperature should remain between 70° and 75°F. Use red aquarium bulbs to achieve this, if necessary. Use a full-spectrum fluorescent bulb over the vivarium for 10 to 12 hours each day, even though this turtle does seem to be somewhat nocturnally active.

Offer the Chinese four-eyed turtle fresh food 3 to 4 times each week. The food can consist of commercial frozen turtle food, stick turtle food, or pellet turtle food, crickets and feeder fish. Earthworms, shrimp and new-born mice can be used as supplements. Offer your turtle some soft fruit (like melon) or leafy greens (like romaine or escarole) because each specimen has its own particular taste, and a varied diet is best.

The Chinese four-eyed turtle is an active turtle. Females tend to be shyer than males, but once in captivity they usually become quite bold and they will learn to accept food from your hand. In fact, they can become so enthusiastic that they will swim towards you as you approach their tank.

Use either a submersible canister filter or an off-the-back filter with an extension tube (as discussed with the Amboina box turtle) to keep the water clean.

Although it is not always available, the Chinese four-eyed turtle is worth the search—it becomes quite friendly and does extremely well in captivity if you give it the proper care.

Yellow-headed box turtle

The yellow-headed box turtle, *Cuora flavomarginata* is also sometimes known as the Chinese box turtle, the snake-eating turtle, or the yellow-margined box turtle. It is native to southeast Asia and China, and, as its name implies, it has a hinged plastron which allows it to close itself up in its shell for self protection.

This turtle has an average plastron length of about 5 inches (Illus. 41). It typically has a brown to olive-colored shell with some scales bordered in yellow. The head is yellow as well. The top of some specimens' heads is an olive color. The top of the shell (the carapace) usually has a yellow ridge (known as a keel) running down the middle.

The yellow-headed box turtle inhabits rice fields and marshy areas in its native habitat. It is primarily a terrestrial animal. It can swim, but not as well as many aquatic turtles, and it usually does not dive below the surface.

This turtle is omnivorous; it eats vegetable and fruit matter, carrion and live animals, including insects and earthworms. It uses its beak and front claws to tear off bite-sized pieces of food. It does not need to feed in the water.

The origin of one of its common names, the snake-eating turtle, is not certain. This name is used in the Orient, and it could come from the sight of this turtle feeding on a large earthworm, a small living snake, or a dead snake. The docile nature of this turtle suggests that it probably would not try to attack a very large living snake.

This is another reptile that is not always recognized for its value as a pet. The yellow-headed box turtle becomes quite tame. It maintains a healthy appetite and it is extremely hardy. In many ways, the yellow-headed box turtle does better in captivity than the American box turtles of the genus *Terrapene* that are so often

Illus. 41. *The yellow-headed box turtle,* Cuora flavomargi-nata, *can close itself completely within its shell when attacked.*

found in pet stores. The yellow-headed box turtle does not seem to be as susceptible to the eye and internal infections that the American box turtles get by the time they reach the pet store.

A 40-gallon aquarium (or equivalent enclosure) is recommended for these turtles because they like to move around quite a bit. The top of whatever enclosure you use should be covered securely to protect the turtles and to prevent escapes. If a dog takes one of these, or any other turtles in its mouth, the resulting wounds could kill the turtle.

A substrate of indoor/outdoor carpeting, newspaper, bark mulch or coarse aquarium gravel is suitable. Do not use sand, or gravel that is too fine. If the turtle eats it, the result could be intestinal blockage.

A large, flat water bowl with ramps on the inside and outside for easy access is recommended. In case your specimen turns out to be a really bad swimmer, maintain the water level in the dish at around 2 inches.

The yellow-headed box turtle likes to bask in the sunlight. Like many reptiles, it absorbs vitamin D_3 in this manner. This vitamin is vital to this animal so that it can metabolize calcium. With this in mind, provide full-spectrum fluorescent lighting for 12 hours each day.

Take the yellow-headed box turtle outside occasionally for some natural sunlight. If you do this, never leave the turtle in a glass or plastic container exposed to the sun, and always make sure that there is adequate shade. The turtle must be able to avoid overheating, or it could die.

If you take your yellow-headed box turtle outside, keep in mind that it is surprisingly stealthy. If you do not watch it closely at all times, it could slip off into any nearby brush, where it may be impossible to find.

Maintain the temperature for the yellow-headed box turtle between 75° and 80°F. Use a red spot bulb focused into one corner of the tank to provide the proper temperature. Use a thermometer to monitor the temperature and to prevent overheating.

The yellow-headed box turtle is omnivorous. It will eat crickets, earthworms, snails (beware of snails contaminated by various poisons), fish, fruits, vegetables and dog food. Offer your turtle fresh food 3 times each week. Provide a variety of foods to ensure proper nutrition for your turtle.

Powdered vitamins for birds or reptiles that contain vitamin D_3 and calcium are a must. Sprinkle the vitamins lightly over the food each time you feed your pet, and you will further guarantee its health.

Weight is one good indicator of this or any other turtle's health. When you pick up a turtle, it should feel heavy, relative to its size. If the turtle is surprisingly light, it could be a sign of a parasitic infection. Take a stool sample to your veterinarian as quickly as possible so that he or she can administer treatment. If medicine is given soon enough, most turtles will recover from a parasitic infection surprisingly well.

Yellow-headed box turtles make excellent reptile pets. They usually become quite tame, and many specimens will learn to take food from your hand. Some will even walk towards you when they see you approach. If you're looking for a turtle, the yellow-headed box turtle is highly recommended.

Metric Equivalents

INCHES TO MILLIMETRES AND CENTIMETRES

MM—millimetres   *CM—centimetres*

Inches	MM	CM	Inches	CM	Inches	CM
⅛	3	0.3	9	22.9	30	76.2
¼	6	0.6	10	25.4	31	78.7
⅜	10	1.0	11	27.9	32	81.3
½	13	1.3	12	30.5	33	83.8
⅝	16	1.6	13	33.0	34	86.4
¾	19	1.9	14	35.6	35	88.9
⅞	22	2.2	15	38.1	36	91.4
1	25	2.5	16	40.6	37	94.0
1¼	32	3.2	17	43.2	38	96.5
1½	38	3.8	18	45.7	39	99.1
1¾	44	4.4	19	48.3	40	101.6
2	51	5.1	20	50.8	41	104.1
2½	64	6.4	21	53.3	42	106.7
2	76	7.6	22	55.9	43	109.2
3½	89	8.9	23	58.4	44	111.8
4	102	10.2	24	61.0	45	114.3
4½	114	11.4	25	63.5	46	116.8
5	127	12.7	26	66.0	47	119.4
6	152	15.2	27	68.6	48	121.9
7	178	17.8	28	71.1	49	124.5
8	203	20.3	29	73.7	50	127.0

About the Author

Jef Hewitt first became fascinated with unusual animals when, at age five, he received a pet lizard from his parents. Now a naturalist and an accomplished wildlife photographer, he frequently speaks at schools and to professional organizations about the more unusual aspects of nature. He lives in Los Angeles with his wife and son, and is the host of the cable television program, "The Wild Side."

INDEX